AF601440

Green Energy and Technology

Climate change, environmental impact and the limited natural resources urge scientific research and novel technical solutions. The monograph series Green Energy and Technology serves as a publishing platform for scientific and technological approaches to "green"—i.e. environmentally friendly and sustainable—technologies. While a focus lies on energy and power supply, it also covers "green" solutions in industrial engineering and engineering design. Green Energy and Technology addresses researchers, advanced students, technical consultants as well as decision makers in industries and politics. Hence, the level of presentation spans from instructional to highly technical.

Indexed in Scopus.
Indexed in Ei Compendex.

Djamila Rekioua

Energy Storage for Renewable Energy Systems

Modeling, Control, and AI Integration

Springer

Djamila Rekioua
UNIVERSITY OF BEJAIA-ALGERIA
Bejaia, Algeria

ISSN 1865-3529 ISSN 1865-3537 (electronic)
Green Energy and Technology
ISBN 978-3-032-19588-3 ISBN 978-3-032-19589-0 (eBook)
https://doi.org/10.1007/978-3-032-19589-0

This Springer imprint is published by the registered company Springer Nature Switzerland AG
The registered company address is: Gewerbestrasse 11, 6330 Cham, Switzerland

Foreword

We stand at a pivotal moment in the global energy transition. The rapid rise of renewable energy sources such as solar and wind has been nothing short of revolutionary, promising a future of cleaner, more sustainable power. Yet, this promise is inherently coupled with a fundamental challenge: its variability. The sun does not always shine, nor does the wind always blow. To bridge the gap between intermittent generation and the constant, unwavering demand for electricity, we have a critical linchpin: energy storage.

It is precisely at this intersection of challenge and opportunity that Prof. REKIOUA's timely and essential book, *Energy Storage for Renewable Energy Systems*: *Modeling, Control, and AI Integration*, makes its mark. This book moves beyond a mere description of battery chemistries or pumped-hydro facilities. It presents a holistic and integrated framework for understanding, modeling, and intelligently managing storage systems as the dynamic heart of the modern renewable-based grid.

This book's structure logically guides the reader through this complex landscape. It begins with a rigorous foundation in the modeling of various storage technologies, a crucial first step for any meaningful analysis. It then progresses to the critical layer of control systems, which govern the real-time charging and discharging cycles to maintain stability and efficiency. Finally, and most innovatively, this book delves into the transformative potential of artificial intelligence. It demonstrates how machine learning and advanced algorithms are not just incremental improvements but are fundamentally reshaping our ability to forecast, optimize, and autonomously control storage assets at scale.

Prof. Djamila REKIOUA has masterfully synthesized a vast body of knowledge, drawing on power systems engineering, control theory, and power electronics. The result is a work that is both academically rigorous and intensely practical. It is an indispensable resource for students embarking on a career in this field, for researchers seeking the frontiers of knowledge, and for practicing engineers tasked with building the resilient and intelligent energy systems of tomorrow.

I have had the pleasure of knowing Dr. Djamila REKIOUA for many years and have always admired her clarity of thought and dedication to advancing our field. In these pages, that dedication is evident on every page. This book is a significant contribution to the literature and a powerful tool for anyone committed to turning the vision of a fully renewable-powered world into a reality.

Swinburne University of Technology
Melbourne, VIC, Australia
November 23rd, 2025

Saad Mekhilef

Acknowledgments Without the invaluable assistance and efforts of many individuals, this book could not have been completed. I am deeply grateful to all those who have provided their support.

I would especially like to thank my colleagues and Ph.D. students in the fields of renewable energy and power electronics at the LTII Laboratory, University of Bejaia, for their collaboration, expertise, and commitment across numerous projects.

My sincere gratitude goes to my family, my husband **Toufik**, and my children **Yasmine**, **Samy**, and **Camelia**, for their constant encouragement, patience, and love. I also extend profound thanks to my parents, whose principles and values have been a continuous source of motivation, determination, and inspiration.

Finally, I express my sincere appreciation to Springer-Verlag for their support in bringing this work to publication.

Djamila Rekioua

Competing Interests The author has no competing interests to declare that are relevant to the content of this manuscript.

Introduction

In renewable energy systems (RESs), energy storage systems (ESSs) play a pivotal role in ensuring reliability, efficiency, and grid stability, particularly due to the intermittent nature of renewable resources. This book delves into both single-storage and multi-storage systems within the context of RESs. The choice between these storage architectures depends on various factors, including grid requirements, load profiles, economic constraints, and the nature of the renewable energy source. As renewable energy continues to expand globally, intelligent storage design and control have become essential to ensure optimal integration into power systems. Through eight well-structured chapters, this book offers a comprehensive and multidisciplinary exploration of ESSs in RESs. It extends from fundamental concepts and modeling techniques to more advanced strategies, making it a valuable resource for those studying smart grid and renewable energy technologies.

Aims of the Book

Although many books address ESSs and their integration into RESs, this one distinguishes itself by combining theoretical foundations with practical methodologies aimed at improving the design, control, and optimization of energy storage within renewable systems. It is structured into different sections, each addressing different aspects of ESS and its applications. The overarching aim is to advance understanding in this field by integrating theoretical insights with practical applications. This book aims to improve the understanding and covers a wide range of ESS-related topics, from fundamental concepts to advanced applications, and by offering a detailed analysis of how energy storage technologies can be effectively optimized and easily incorporated into RESs.

How the Book Is Organized?

This book provides a methodical and comprehensive overview of ESSs used in RESs, including both the theoretical basis and practical applications. This book leads the reader through eight focused chapters from basic storage concepts to intelligent control techniques, including artificial intelligence (AI) for optimizing energy systems.

The book is divided into eight chapters:

Chapter 1 introduces the essential role of ESSs in RESs, highlighting the factors that make storage indispensable for managing renewable variability and ensuring system reliability.

Chapter 2 provides an overview of the ESSs commonly used in RESs, emphasizing the parameters that influence the selection, design, and operation of storage technologies.

Chapter 3 presents single-storage and multi-storage systems for RESs. It discusses the various storage technologies used alongside PV and wind turbine systems and explains the operational synergies that arise from combining multiple storage devices.

Chapter 4 examines storage integration methods in hybrid energy systems, focusing on design aspects, efficiency considerations, and practical applications.

Chapter 5 presents modeling techniques used to simulate, analyze, and optimize ESSs. Accurate modeling is essential for assessing system performance under diverse operating conditions.

Chapter 6 addresses energy management strategies for multi-storage systems in PV, wind, and hybrid configurations, covering advanced control algorithms, real-time optimization methods, and practical implementation aspects.

Chapter 7 explores optimization and integration strategies for ESSs within RESs, supported by practical examples demonstrating how optimization enhances system reliability and reduces operating costs.

Finally, Chap. 8 presents applications of ESSs in RESs using MATLAB/Simulink. It includes case studies for PV, wind, and hybrid systems, demonstrating modeling, simulation, and control approaches.

Notations

General

v, **V**	Voltage in instantaneous and vector notation
i, **I**	Current in instantaneous and vector notation

Subscripts

d, q	Quantities in d-axis and q-axis
α,β	Quantities in α-axis and β-axis
s, r	Stator and rotor
a, b, c	Quantities in phases a, b, and c

Symbols

α	Weighting factor in SOC–OCV fusion
$\alpha_0, \alpha_1, \alpha_2, \alpha_3$	Polynomial coefficients for OCV–SOC curve
$\alpha_{n1}, \alpha_{n2}, \alpha_{n3}$	Empirical coefficients in modified Nernst expression
β_i	Empirical fitting coefficients
Δf	Frequency deviation
Δf_d	Deadband frequency threshold
ΔG	Gibbs free energy change
ΔP	Power variation
ΔP_{pv}	PV power variation after perturbation
ΔP_{Tb}	Wind turbine power variation after perturbation
Δt	Time step duration
η	Round-trip efficiency

η_{batt} Battery efficiency

η_{FC} Fuel cell efficiency

η_{losses} Photovoltaic loss factor

η_{charge} Charging efficiency

$\eta_{discharge}$ Discharging efficiency

η_{mech} Mechanical efficiency

η_{pump} Pump efficiency

η_{rt} Round-trip efficiency

η_{turb} Turbine efficiency

η_{sto} Round-trip storage efficiency

λ Membrane water content

ρ_{air} Air density

σ_m Proton conductivity

ω Rotational speed

Ω_{FESS} Flywheel angular speed

General Parameters

A_f Vehicle frontal surface area

C_1, C_2 Capacitors in EC model branches

C_{capex} Capital expenditure cost

C_d Aerodynamic drag coefficient

C_{dc} dc bus capacitance

C_{deg} Battery degradation cost factor

C_{gen} Cost coefficient for generation

C_{inv} Initial investment cost

C_k Capacitance of RC branch

C_{nom} Nominal capacity

$C_{O\&M}$ Operating and maintenance cost over lifetime

C_{opex} Operational and maintenance cost

C_t Operational cost at time t

C_{stor} Cost coefficient for storage usage

D Battery degradation rate

D_{cycle} Duty cycle of DC–DC converter

D_{pv} Duty cycle of DC–DC converter in PV system

D_{Tb} Duty cycle of DC–DC converter in wind turbine system

$E_{available}$ Available energy

E_{batcap} Battery energy capacity

$E_{batt,max}$ Maximum battery capacity

$E_{batt,nom}$ Nominal battery energy

$E_{batt,req}$ Required battery energy

E_{cell} Energy per SC cell

E_{def}	Energy deficit
E_{daily}	Daily energy demand
$E_{delivered}$	Total delivered energy
E_{in}	Input energy during charging
E_{FESS}	Flywheel usable energy
E_{H2}	Hydrogen energy required
E_{max}	Maximum energy capacity
E_{OCV}	Open-circuit voltage function
E_{out}	Output energy during discharge
E_{PV}	Photovoltaic daily energy
E_{sc}	Supercapacitor pack energy
E_{s}	Solar irradiance
$E_{supplied}$	Energy supplied to battery during charge
E_{sto}	Stored energy
E_{total}	Total daily energy demand
$e(t)$	Frequency error
F	Faraday constant
F_{aero}	Aerodynamic drag force
F_{r}	Total resistive force
F_{ro}	Rolling resistance coefficient
F_{slope}	Climbing force
F_{tire}	Rolling resistance force
$f(c)$	Objective function for hybrid combinations
$f(x)$	Objective function to minimize
$f_{sharing}$	Sharing factor
$g_{i}(x)$	Inequality constraints
$h_{j}(x)$	Equality constraints
H_{virt}	Virtual inertia constant
I_{0}	Exchange current
I_{batt}	Battery current
$I_{batt-ref}$	Reference battery current
I_{ch}	Charging current
I_{dis}	Discharging current
I_{h}	Hourly irradiance fraction
I_{lim}	Limiting current
I_{max}	Peak current of SC pack
I_{RDC}	Standard discharge current
j	Current density
J_{g}	Generator inertia
$J_{inertia}$	Moment inertia
J_{Tb}	Turbine inertia
k	Empirical constant related to battery chemistry
K_{droop}	Droop coefficient
K_{i}	Integral gain
K_{p}	Proportional gain

K_s	Secondary control gain
L_{fc}	Boost inductor
L_{sc}	Buck–boost inductor
l_m	Membrane thickness
m_{H2}	Hydrogen mass
N	Lifetime
N_0, N_1	Cycle life indicators
N_{batt}	Number of batteries
N_{cycles}	Number of cycles
P	Power rating
P_{batt}	Power of battery energy storage
P_{charge}	Charging power
$P_{discharge}$	Discharging power
P_{ESS}	Power of ESS
P_{FC}	Fuel cell electrical output
P_{FCmax}/P_{FCmin}	Fuel cell power limits
P_{gen}	Generated power
P_{load}	Power demand
$P_{grid\text{-}imp}$, $P_{grid\text{-}exp}$	Grid import and export power
P_{loss}	Power losses
P_{peak}	Peak demand
P_{pv}	Photovoltaic power
$P_{PV,peak}$	Photovoltaic peak power
P_{rated}	Rated power output
$P_{reduced}$	Peak reduced by ESS
P_{ref}	Reference power
P_{wind}	Wind power
$Q_{remaining}$	Remaining charge
$Q_{nominal}$	Nominal capacity
$R_{c,area}$	Contact resistance
R_{gas}	Universal gas constant
R_f	Leakage resistance
R_i	Initial resistance (electrolyzer)
R_{int}	Internal resistance
R_k	Polarization resistance
R_o	Ohmic/internal resistance
R_{stator}	Stator resistance
R_s	Equivalent series resistance (supercapacitor)
$R_{m,area}$	Membrane resistance
R_{max}	Maximum ramp rate
r	Discount rate
$s_1 \ldots s_4$	Storage technologies in combination examples
t	Time
t_d	Demand time
$t_{discharge}$	Time to fully discharge

t_g	Generation time
t_{supply}	Flywheel supply time
$T_{°C}$	Temperature in °C
T_{FC}	Fuel cell temperature
T_{peak}	Duration of peak load
T_{PEMFC}	PEMFC temperature
V	Voltage
V_{actual}	Actual voltage
V_b	Reservoir volume
V_{ch}	Charging voltage
V_{dc}	dc bus voltage
V_{dis}	Discharging voltage
V_{FC}	Fuel cell voltage
V_{H2}	Hydrogen production volume
V_i	Ideal electrochemical potential
V_m	Molar volume
V_{max}	Supercapacitor maximum voltage
V_{pres}	Reservoir pressure
V_{ref}	Reference voltage
V_{series}	Supercapacitor series voltage target
V_{sc}	Supercapacitor voltage
V_N	Supercapacitor nominal voltage
v_1, v_2	Capacitor voltages in RC branches
x	Vector of decision variables

Superscripts

AC	Alternating current
AI	Artificial intelligence
ANFIS	Adaptive neuro-fuzzy inference system
ANN	Artificial neural networks
BESS	Battery energy storage system
BES	Battery energy storage
CAES	Compressed air energy storage
CES	Capacitor energy storage
DOD	Depth of discharge (%)
DOD_{actual}	Actual operating depth of discharge
EC	Energy capacity (classification criterion)
EI	Environmental impact (classification criterion)
EMS	Energy management system
ESS	Energy storage system
FES	Flywheel energy storage
FLC	Fuzzy logic controller

FOC Field oriented control
H_2 Hydrogen
LA Lead-acid storage
LCOS Levelized cost of storage
LHVH2 Lower heating value of hydrogen
Li-ion Lithium ion
LP Linear programming
LT Long term
M1–M8 Operating modes in management algorithms
MPC Model predictive control
MPPT Maximum power point tracking
MS Mechanical storage
MT Medium term
NN Neural network
NPV Net present value
OTC Optimal torque control
P&O Perturb and observe
PEMFC Proton exchange membrane fuel cell
PHES Pumped hydro energy storage
PI Proportional integral
PMC Power management control
PMSG Permanent magnet synchronous generator
PSH Peak sun hours
PV Photovoltaic
PWM Pulse-width modulation
RE Renewable energy
RES Renewable energy system
SCES Supercapacitor energy storage
SMES Superconducting magnetic energy storage
ST Short term
SOC State of charge
SOC_{max} Maximum allowed SOC
SOC_{min} Minimum allowed SOC
SOH State of health
TES Thermal energy storage
VST Very short term
WTbs Wind turbine-based systems

Contents

Chapter 1
Necessity of Storage in Renewable Energy Systems

1.1 Introduction

This chapter examines the ESSs and their importance in energy systems and especially in renewable energy systems (RESs). It establishes the foundation by giving the essential elements that make energy storage systems (ESSs) necessary for supporting intermittent renewable energy sources. Acting as a fundamental reference, this chapter is particularly valuable for researchers aiming to fully understand the essential role that storage plays in RESs. The importance of ESSs in RESs is discussed at length, with important elements identified.

1.2 Understanding Energy Storage

Understanding the different types of storage systems and their uses is important. Because they store energy through electrochemical processes, batteries for example are perfect for grids and electric vehicles. Using gravity, pumped hydro stores energy by pumping water to a higher elevation and releasing it when required [1]. By rotating a rotor, flywheels store kinetic energy and provide high cycle life, and fast response. Energy storage technologies operate on a few fundamental principles, such as charging, storing, and discharging energy. Their operation differs based on capacity, efficiency, and suitability for different applications. For example, pumped hydro storage excels in large-scale, long-duration storage, while batteries are more efficient and flexible for small-scale or decentralized applications. Flywheels are often used where fast and frequent energy discharge is required. By understanding these systems, we can assess their roles in stabilizing the grid and in RESs [2–5].

D. Rekioua, *Energy Storage for Renewable Energy Systems*, Green Energy and Technology, https://doi.org/10.1007/978-3-032-19589-0_1

1.3 Storage Importance in RESs

Necessity of storage in RESs is driven by several fundamental factors due mainly to the nature of renewable, grid stability, and energy reliability.

1.3.1 Renewable Energy Sources Nature

Storage in RESs is necessary because of the irregular nature of RE resources. These sources are weather-dependent, and their energy output varies during the day and throughout seasons (Fig. 1.1). In fact, the generation of renewable energy is susceptible to fluctuations due to various variables, and climatic conditions [6–10].

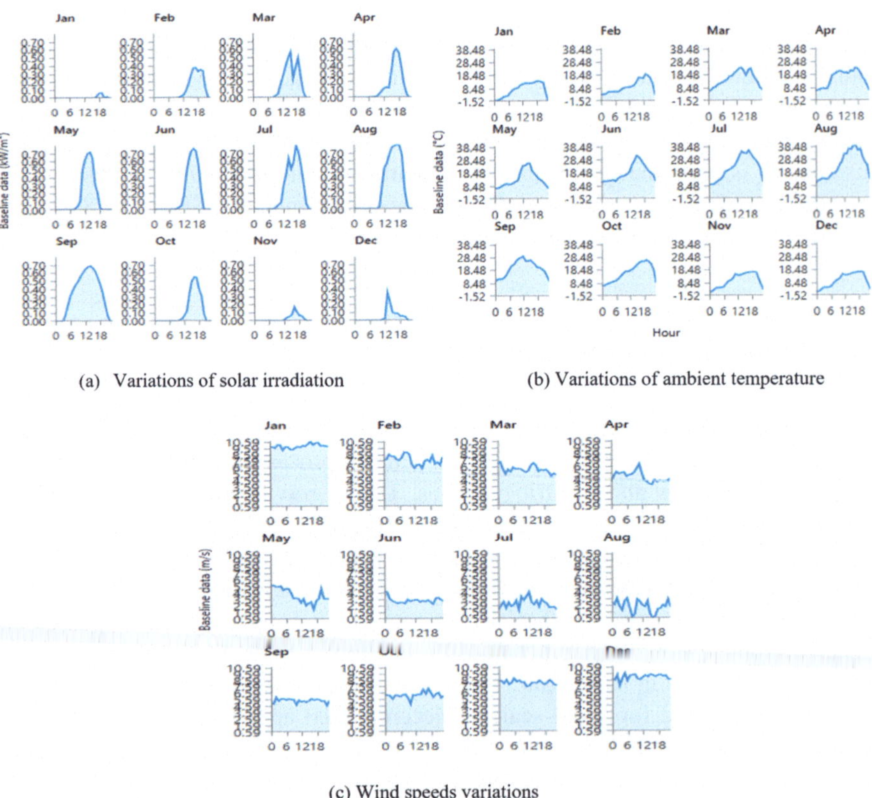

(a) Variations of solar irradiation

(b) Variations of ambient temperature

(c) Wind speeds variations

Fig. 1.1 Variations in weather conditions. (**a**) Variations of solar irradiation. (**b**) Variations of ambient temperature. (**c**) Wind speeds variations

1.3.2 Grid Stability

Grid stability is enhanced by energy storage devices, which act as a regulator against abrupt fluctuations in the production of renewable energy [11, 13] (Fig. 1.2).

1.3.2.1 Fast-Response Solutions for Frequency Regulation

Storage systems give fast-response solutions for frequency regulation to stabilize the grid. The overall system power balance is [11–14]:

$$P_{gen}(t) + P_{batt}(t) = P_{load}(t) \tag{1.1}$$

1.3.2.2 Basic Control Methods

The basic control methods for frequency control typically involve the following strategies:

1.3.2.2.1 Droop Control

It is among the most common methods used in frequency regulation. The principle involves adjusting BESS output power proportionally to the deviation from the nominal grid frequency [11–17].

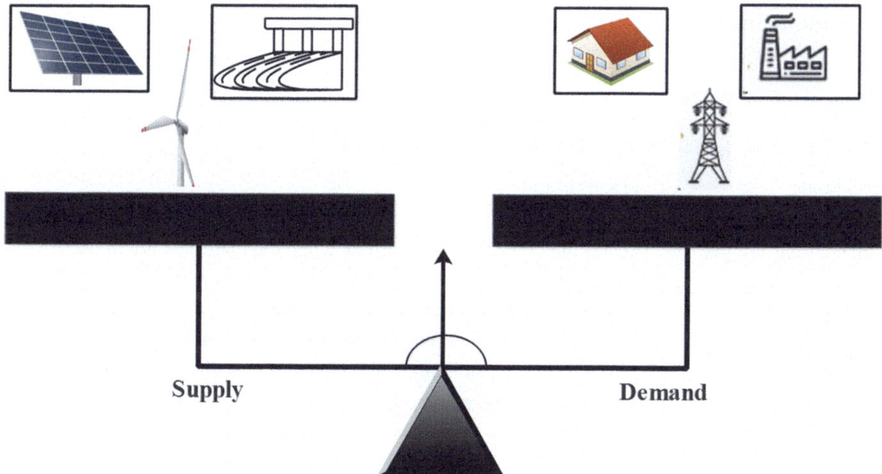

Fig. 1.2 Electrical system model with storage

$$P_{batt}(t) = -K_{droop}(f \_ f_{nom}) \tag{1.2}$$

When the frequency falls below a specific level, the battery discharges the load demand, and if the frequency exceeds the specified limit (indicating excess generation), the battery absorbs power. Droop control provides a decentralized and autonomous response, making it widely applicable.

1.3.2.2.2 Proportional Integral Control (PI)

PI control uses a control loop to modify the battery's power output. The proportional part reacts to the immediate frequency deviation, while the integral part accounts for the accumulated deviation over time. This combination allows for a more precise and stable frequency response, particularly useful in mitigating frequency drifts over longer periods [11].

$$\begin{gathered} P_{batt} = K_p \,.\, e(t) + K_i \,.\, \int_0^t e(t) \,.\, dt \\ e(t) = f_{nom} - f \end{gathered} \tag{1.3}$$

1.3.2.2.3 State of Charge (SOC) Estimation Method

This method aims to determine the actual SOC of the battery. It uses this diagnostic or computational procedure to measure the battery energy. The widely used methods are given in Table 1.1. To calculate the SOC, which is crucial for the internal control of battery systems, they rely on primary input data such as current, voltage, temperature, and time.

Table 1.1 Main used techniques in SOC estimation

Technique	Description
Coulomb counting	Estimates SOC
Open circuit voltage (OCV)	Uses the voltage-SOC relations when the battery is not in use.
Kalman filter	Combines measurements and the model to ensure accuracy
Extended Kalman filter	Uses nonlinear version of Kalman filter for better accuracy
Neural networks	Uses data to learn SOC while adjusting to battery performance
Adaptive observers	Adjusts model parameters online in accordance with changing conditions

The SOC is expressed as:

$$SOC(t+1) = SOC(t) + \frac{1}{C_{nom}} \int_{t}^{t+1} I_{batt}(t) \, . \, dt \tag{1.4}$$

Example:

The SOC estimate is frequently adjusted using open-circuit voltage (OCV) feedback to compensate for integration error and current sensor offset that are present in the Coulomb counting method. This adjustment provides the battery state estimation's accuracy. Battery characterization studies done under controlled temperature and current conditions are used to experimentally assess the relationship between OCV and SOC.

A third-order polynomial function is typically used to express it:

$$SOC_{OCV} = \alpha_0 + \alpha_1 \, . \, SOC + \alpha_2 \, . \, SOC^2 + \alpha_3 \, . \, SOC^3 \tag{1.5}$$

The Coulomb-counting estimate and the OCV-based estimate are then combined by averaging them to determine the corrected SOC value:

$$SOC_{corrected} = \alpha \, . \, SOC_{coulomb} + (1-\alpha) \, . \, SOC_{OCV} \tag{1.6}$$

The weighting factor α, which is usually between 0.7 and 0.9, establishes a balance between long-term stability (OCV correction) and responsiveness (Coulomb estimation).

1.3.2.2.4 State of Charge (SOC) Management

The control method must balance the need to provide frequency support while maintaining the SOC within an optimal range. For example, when SOC is low, the system may limit discharging to preserve battery life and reserve capacity for critical responses. Conversely, when SOC is high, the system might prioritize charging or frequency support [16–24].

We have:

- If $SOC < SOC_{min}$ No discharge allowed
- If $SOC > SOC_{max}$ No charge allowed

Thus:

$$SOC(t+1) = SOC(t) + \frac{P_{batt} \, . \, \Delta t}{E_{\max}} \tag{1.7}$$

The immediate power balance between generation and demand determines how it operates. In order to avoid both overcharging and severe discharge, as the battery's

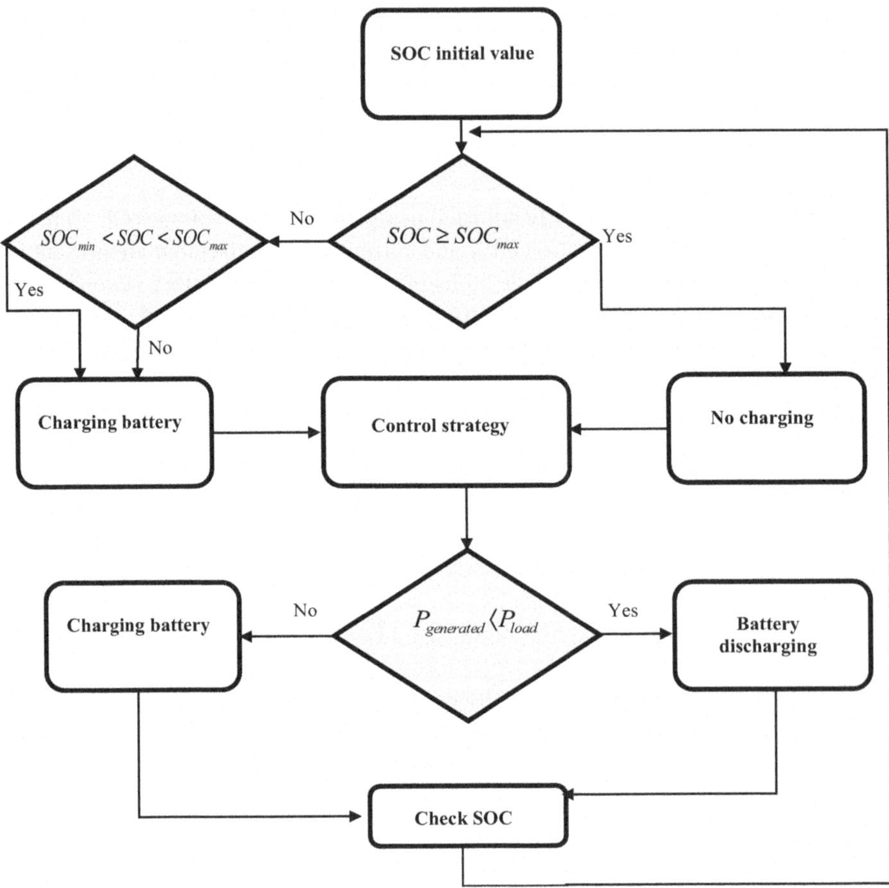

Fig. 1.3 Flowchart for charging and discharging a battery

SOC gets closer to its lower limit, the overall control technique makes sure that it stays within the predetermined limits, the control process initiates charging to maintain it above this level. In the same manner, discharging is allowed to satisfy load demand when the SOC exceeds its upper limit, but charging is limited to keep the SOC from exceeding above this higher limit. A battery charge–discharge management logic is described in Fig. 1.3.

1.3.2.2.5 Deadband Control

It introduces a range around the nominal frequency where the BESS does not react to small frequency deviations. This method avoids unnecessary cycling and prolongs the storage system's life. The BESS only activates when the frequency deviation exceeds the predefined deadband and ensuring responses are reserved for significant imbalances [16, 17, 21–25].

$$P_{batt} = -K_{droop}\left(f - f_{nom}\right)\ if\ \ \left|f - f_{nom}\right| > \Delta f_d \tag{1.8}$$

$$P_{batt} = 0 \quad otherwise \tag{1.9}$$

With: Δf_d is the deadband

1.3.2.2.6 Virtual Inertia Control

By using BESS to instantly absorb or inject power in response to frequency changes, it replicates the behavior of conventional synchronous generators. By providing synthetic inertia, BESS helps stabilize the grid during sudden disturbances [19].

$$P_{batt} = -H_{virt} \cdot \frac{df(t)}{dt} \tag{1.10}$$

1.3.2.2.7 Droop-Free Control

Unlike droop control, droop-free or secondary control methods aim to restore frequency to its nominal value. This method typically involves a centralized controller that coordinates the actions of multiple energy storage systems, adjusting their output to gradually correct the frequency after the initial response provided by primary control methods like droop [17, 18].

$$\frac{df(t)}{dt} = K_s \,.\, \Delta f(t) \tag{1.11}$$

1.3.2.2.8 Ramp Rate Control (RRC)

In scenarios where rapid changes in power output are required, RRC ensures that ESS can change its output smoothly and within limits, preventing sudden spikes or drops in power. This is particularly important for ensuring stability in systems with high penetration of renewables, where frequency deviations can be more volatile [20].

$$\frac{dP_{batt}}{dt} \le R_{\max} \tag{1.12}$$

1.3.2.3 Peak Demand Management

Also, the peak generation times for renewable generation often does not align with peak demand times (e.g., solar power generation peaks in the afternoon, while residential energy use typically peaks in the evening). So, storage systems shift the timing of energy delivery from when it is generated to when it is needed. This load shifting helps meet peak demand without overloading the grid or requiring backup fossil fuel plants [26–29].

To shift energy from generation time to demand time, the total energy stored and later discharged must match and can be written by neglecting losses:

$$E_{stored}\left(t_g\right) = E_{discharged}\left(t_d\right) \tag{1.13}$$

The ESS switches between charging and discharging modes based on generation and demand profiles:

- During charging:

$$P_{ESS}\left(t\right) = -P_{excess}\left(t\right) \quad if \quad P_{gen}\left(t\right) > P_{load}\left(t\right) \tag{1.14}$$

- And during discharging:

$$P_{ESS}\left(t\right) = +P_{required}\left(t\right) \quad if \quad P_{gen}\left(t\right) < P_{load}\left(t\right) \tag{1.15}$$

During peak demand management, the SOC is:

- During charging:

$$\frac{dSOC\left(t\right)}{dt} = \frac{\eta_{charge} \cdot P_{ESS}\left(t\right)}{E_{\max}} \tag{1.16}$$

- And during discharging:

$$\frac{dSOC\left(t\right)}{dt} = -\frac{\eta_{charge} \cdot P_{ESS}\left(t\right)}{\eta_{discharge} \cdot E_{\max}} \tag{1.17}$$

SOC is typically bounded as:

$$SOC_{\min} \le SOC \le SOC_{\max} \tag{1.18}$$

To determine how much the ESS can reduce grid peak demand, it is calculated as:

$$P_{peak}^{new} = P_{peak}^{original} - P_{ESS}\left(t_{peak}\right) \tag{1.19}$$

1.3.2.4 Renewable Energy Penetration

It is crucial to know that as more renewable energy is integrated into the grid, managing variability becomes increasingly difficult without adequate storage. In this case, storage systems enable higher renewable energy penetration by acting as a bridge [6, 7, 13] (Fig. 1.4).

1.3.2.5 Microgrids/off-Grid Applications

In off-grid applications, RE may be the primary or only source of electricity. In these cases, batteries and other storage solutions provide essential backup and stabilization, permitting microgrids to run on RE sources without the assistance of the main electrical system [30–33] (Fig. 1.5).

1.3.3 Energy Reliability

Without storage, renewable energy systems are less reliable due to their dependence on environmental factors. In critical applications such as hospitals and emergency services. So, storage solutions provide backup power during outages, improving reliability [30–33] (Fig. 1.6).

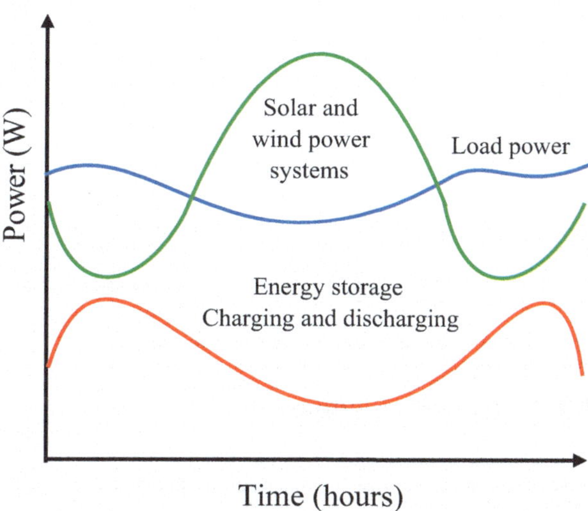

Fig. 1.4 Balancing renewable variability with energy storage

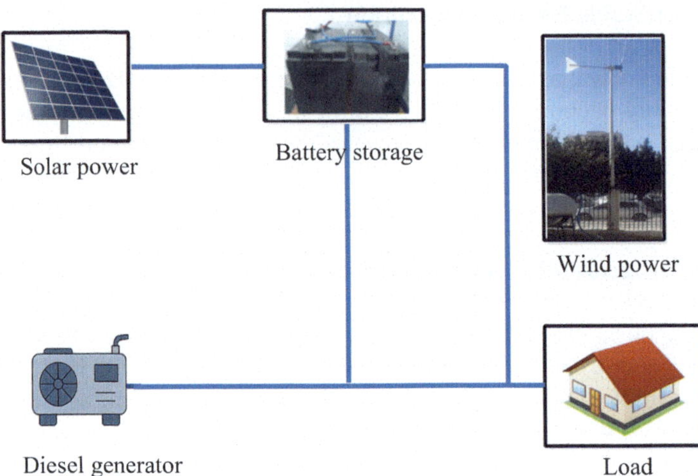

Fig. 1.5 Microgrid with renewable energy and battery storage

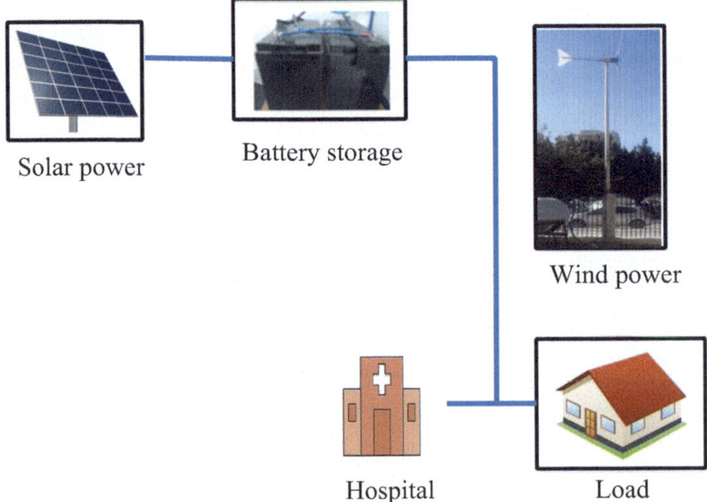

Fig. 1.6 Backup power for reliable renewable energy

1.3.4 Cost Reductions

Advances in ES technologies, such as batteries, are resulting in lower costs, making storage solutions more economically viable for renewable energy projects. Affordable storage options increase the competitiveness of RESs against conventional energy sources [34–37] (Fig. 1.7).

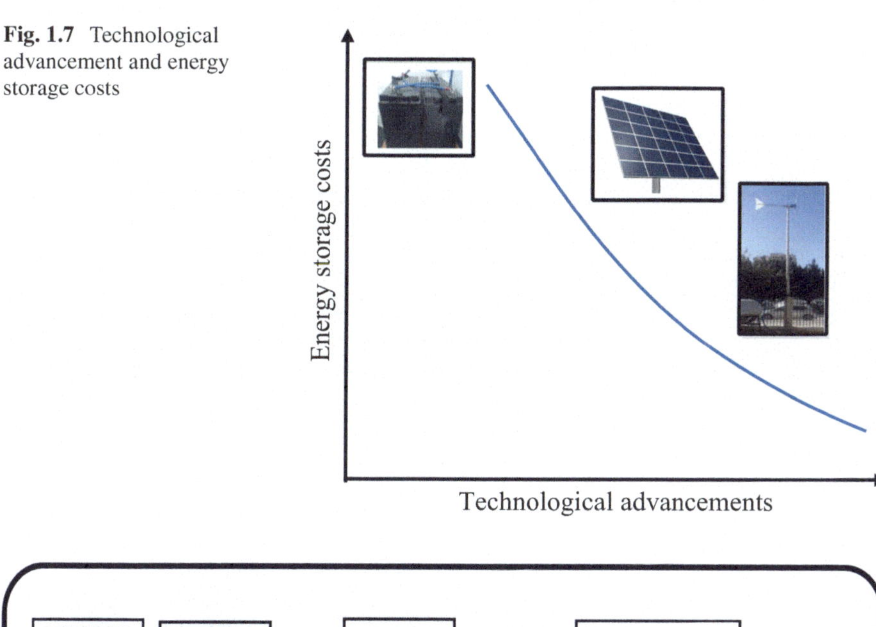

Fig. 1.7 Technological advancement and energy storage costs

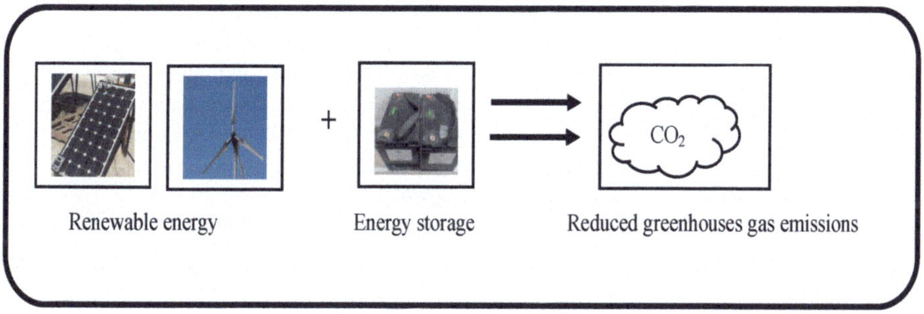

Renewable energy + Energy storage → Reduced greenhouses gas emissions (CO_2)

Fig. 1.8 Environmental benefits

1.3.5 Environmental Benefits

Energy storage supports increased use of RES and contributes to a lower-carbon energy system. In fact, when renewable energy production is low, fossil fuel producers often have to provide backup power in the absence of storage, which decreases the environmental advantages of renewable energies [37, 38] (Fig. 1.8).

1.4 The Different Elements for Developing ESSs

ESS development requires careful consideration of both technical and economic factors. By addressing these considerations, developers can create efficient, reliable, and scalable systems that meet current and future energy storage needs; support

Table 1.2 Main considerations that must be considered when developing ESSs

Consideration	Description
Application and use case	Define the primary purpose
Technology selection	Choose the appropriate technology based on energy, power, and cost requirements
System sizing and capacity planning	Determine the required energy and power capacities for the application to ensure efficient operation
Control strategies and system integration	Implement control systems for SOC management, energy management, and grid service participation.
Economic viability and cost analysis	Assess CapEx, OpEx, revenue streams, ROI, and payback period for financial sustainability and feasibility
Safety and regulatory compliance	Ensure thermal management and environmental regulations
System reliability and longevity	Focus on battery life cycles, fault tolerance
Scalability and modularity	Design systems that are easily extended and adjusted according to varying energy needs and technological advancements
Environmental impact and sustainability	Select sustainable materials, plan for recycling, and conduct lifecycle analysis for a minimal environmental footprint
Advances in materials and technology	Incorporate innovations like solid-state batteries, sodium-ion, and nanotechnology for better performance and safety
Energy efficiency and losses	Optimize round-trip efficiency and minimize conversion losses using efficient power electronics and control systems
Future-proofing and technological innovation	Ensure compatibility with future technologies and allow for upgrades and interoperability

renewable energy integration, and offer sustainable, cost-effective solutions [1, 2, 34, 35, 39, 40].

Table 1.2 provides a structured overview of the basic elements that must be considered when developing effective and future-ready Energy Storage Systems (ESSs).

1.4.1 Technical Challenges in Integrating Energy Storage

Numerous technological difficulties arise when integrating energy storage into renewable energy systems (RES), which may affect the system's overall efficiency [39, 40].

The power equation is:

$$P_{load} = P_{PV} + P_{wind} + P_{batt} \tag{1.20}$$

Figure 1.9 illustrates challenges and integration architecture including renewable sources (PV/Wind), ESS with bidirectional flow, inverter, load, and grid interface.

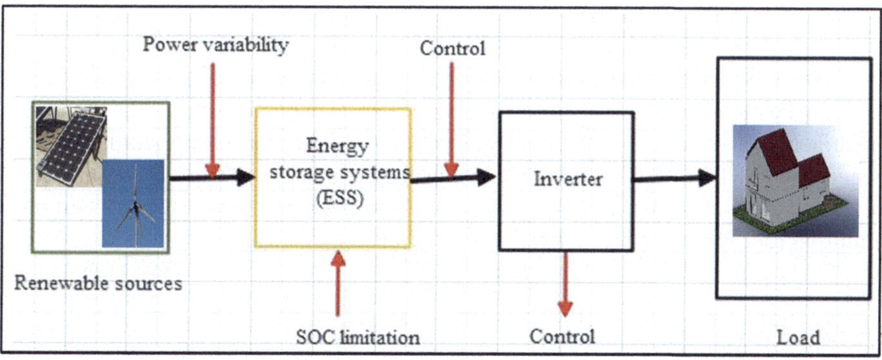

Fig. 1.9 Challenges and integration architecture

1.4.1.1 Capacity Limits

Energy storage systems (ESSs) have limitations in how much energy they can store, which can restrict their effectiveness, particularly when it comes with large-scale renewable energy integration. For example, batteries such as lithium-ion have finite capacity, which may store enough energy during long periods with low solar irradiation or low wind. Scaling up the storage to meet grid-level demands requires more space, materials, and infrastructure [11, 39].

The total energy capacity is:

$$E_{total} = N_{batt} \times C_{batt} \times V_{batt} \tag{1.21}$$

1.4.1.2 Energy Losses

No energy storage technology is perfectly efficient. Both the charging and discharging actions involve energy losses because of different factors like internal resistance in batteries or friction in mechanical storage systems [34, 39, 40].

The storage efficiency $\eta_{storage}$ can be written as:

$$\eta_{storage} = \frac{E_{in}}{E_{out}} \tag{1.22}$$

Or:

$$\eta_{storage} = \eta_{pump} \cdot \eta_{turbine} \cdot \eta_{meca} \tag{1.23}$$

1.4.1.3 Cost of Integration

The cost of integrating energy storage into new or existing energy systems is among the main challenges. The price of ESS technologies can be high due to the materials, manufacturing processes, and installation costs. Additionally, integrating storage systems with RESs requires advanced control systems and infrastructure, upgrades to effectively control the transfer of energy between generation, storage, and consumption [34, 39, 40]. (Fig. 1.10).

1.4.1.4 Energy Density

The energy contained in a given volume or mass is energy density. This limits the practicality of certain storage solutions for grid-scale applications. For example, BESSs require additional space and mass to store the same energy since they have a lower energy density than fossil fuels [11, 34, 39].

1.4.1.5 Degradation and Lifespan

Batteries and other energy storage devices deteriorate with use over time, reducing their capacity to store and release energy effectively. Each charge and discharge cycle leads to some level of degradation, reducing the overall lifespan of the system [21, 23].

Battery capacity degradation over time can be given by:

$$C(t) = C_0 (1 - D)^n \tag{1.24}$$

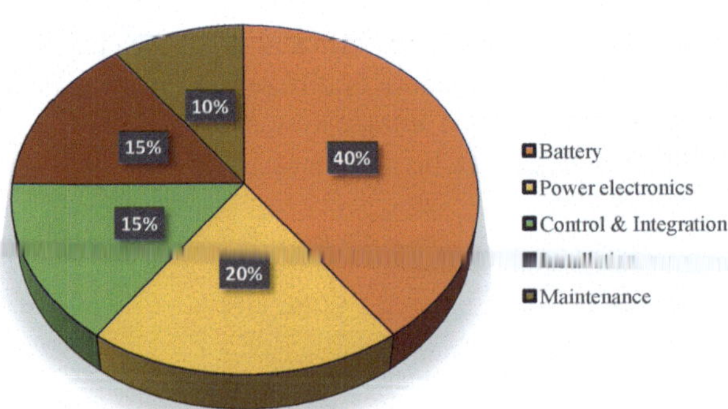

Fig. 1.10 Cost of integrating energy storage

1.4.1.6 Energy Storage in Grid Integration

Integrating ESSs with the existing grid needs advanced control systems to ensure the smooth operation of charging and discharging processes. Coordinating these fluctuations, with energy storage, while maintaining grid stability, involves complex algorithms, real-time monitoring, and prediction systems. Failure to optimize this control can result in inefficiencies, increased costs, and even potential grid instability. Figure 1.11 shows the different inputs as PV & Wind generation and load demand, the Energy Management System (EMS) which monitors generation, *SOC*, and demand in real-time, predicts PV/wind availability using forecasting and makes decisions (charge/discharge/idle), and finally the different outputs for ESSs action (charge/discharge), frequency regulation, voltage support [11, 13, 30].

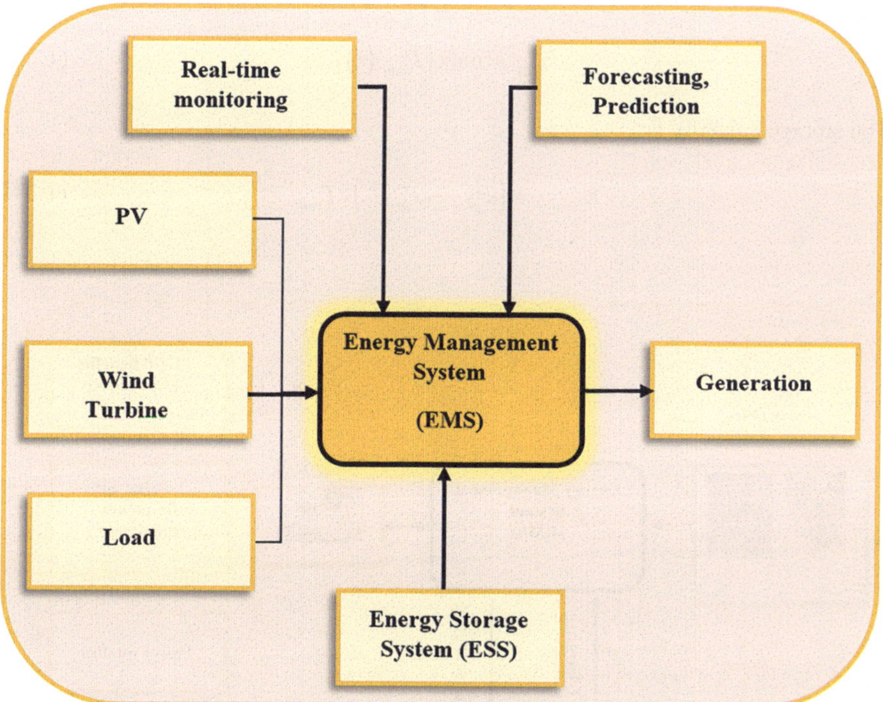

Fig. 1.11 Coordinated control architecture for grid-integrated energy storage systems

1.4.2 Integration of Storage with RESs

Storage must be linked with RESs to optimize energy efficiency and ensure power supply sustainability [1, 3, 30] (Fig. 1.12).

1.4.2.1 Assessing Energy Demand and Load Profiles

Energy demand profiles must be thoroughly examined to optimize the energy storage efficiency. The total daily energy demand is estimated as [26–29]:

$$E_{demand} = \int_{0}^{24} P_{load}(t) \, . \, dt \tag{1.25}$$

The peak load demand (for storage sizing) can be written as:

$$P_{peak} = \max\left(P_{load}(t)\right) \tag{1.26}$$

The storage capacity is:

$$E_{storage} = \left(P_{peak} - P_{gen}^{renew}\right) . \, T_{peak} \tag{1.27}$$

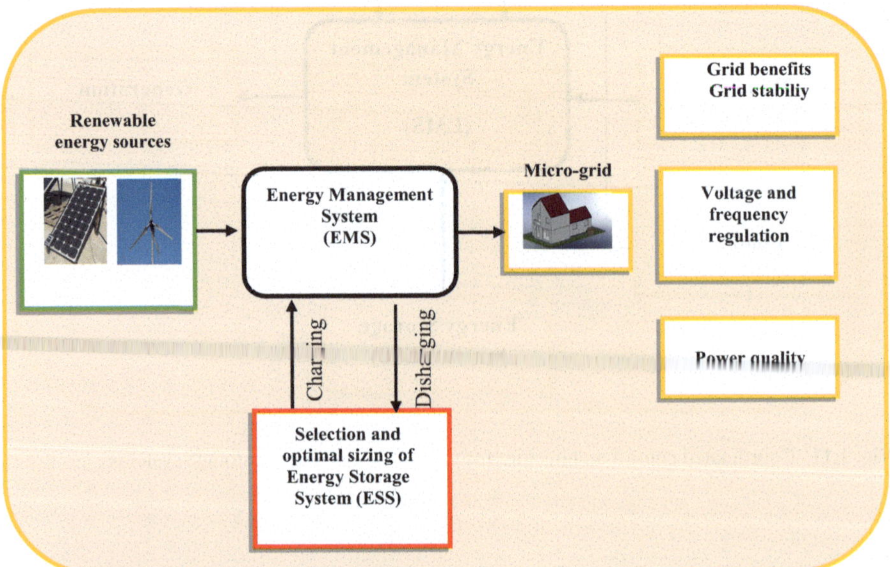

Fig. 1.12 Strategic integration of ESSs into RESs

1.4.2.2 Optimal Sizing of Storage Systems

An undersized storage system may lead to inefficiencies and missed opportunities to store excess energy, while an oversized system could result in unnecessary costs. The storage system should be optimally sized based on factors like local grid demand, potential renewable energy surplus, and expected backup needs (Fig. 1.13).

1.4.2.3 Energy Storage Technology Selection

Different storage technologies are suited for different applications. Pumped hydro systems (PHS) is suitable for long-term, large-scale storage but requires the right geographic circumstances, whereas lithium-ion batteries work well for regular cycling and short-term energy storage (STES). Supercapacitors (SCs) and flywheels (FESS) work well for applications like grid frequency regulation that are suited for quick-response applications [1, 6, 20, 23, 29] (Fig. 1.14).

The coordination of generation, storage, and grid interactions may be optimized by advanced energy management systems (EMS). Based on the expected demand and available generation, EMS can forecast energy production from RE sources and modify the operation of ESSs (Fig. 1.15).

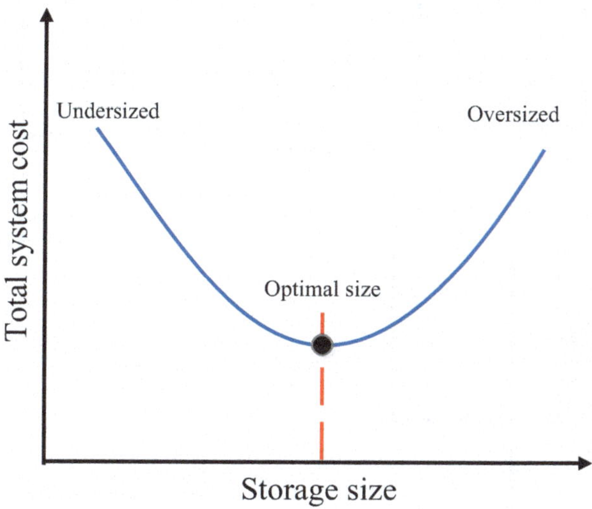

Fig. 1.13 Impact of storage on cost and efficiency

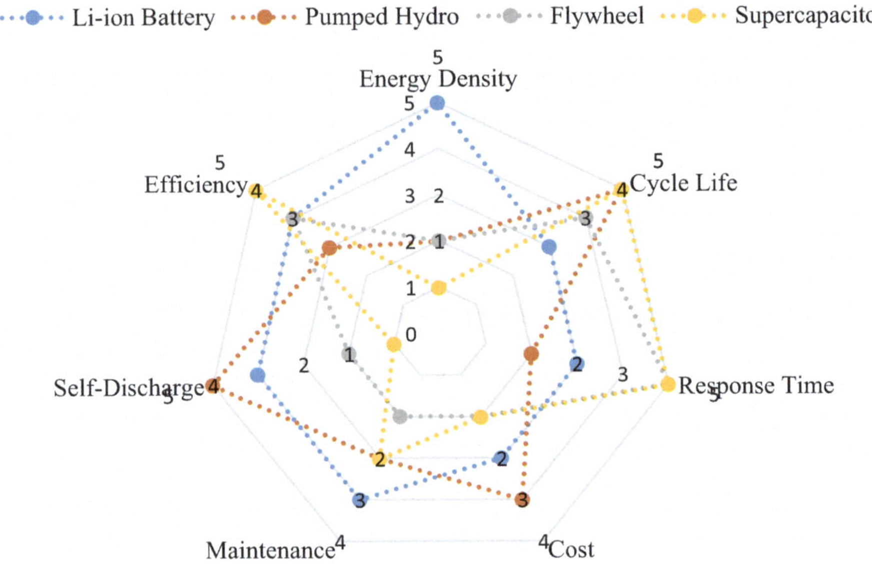

Fig. 1.14 Energy storage comparison

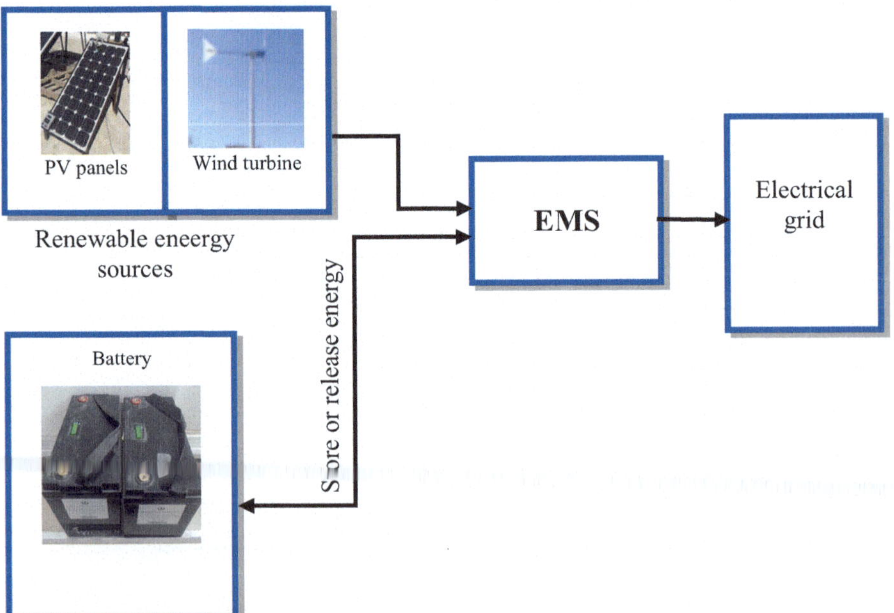

Fig. 1.15 Integration of storage with the grid

1.4.2.4 Geographical and Climatic Considerations

Climate and topography can have an impact on how well energy storage works. For instance, pumped hydro requires specific area, while batteries may perform differently depending on temperature [6, 7, 20]. Table 1.3 gives some examples of recommended storage by geographical region.

1.4.2.5 Grid Services and Ancillary Benefits

By injecting or absorbing electricity in response to variations from the nominal frequency, storage systems helps in maintaining grid frequency. The principle operation is based on power imbalance in the grid:

- If $P_{gen} \neq P_{load}$
  - There is a frequency deviation
- If $P_{gen} < P_{load}$
  - Frequency drops
  - ESS discharges (injects power to the grid
  - Helps in adjusting frequency to its nominal value.
- If $P_{gen} > P_{load}$
  - Frequency rises
  - ESS charges (absorbs excess power)
  - Helps bring frequency back down to nominal

Table 1.3 Recommended storage by geographical region

Region	Recommended Storage	Why?
Mountainous area	Pumped hydro storage	Availability of elevation differences needed for water reservoirs
Deserts	Lithium-ion or thermal storage	High solar irradiance, but extreme temperatures in batteries must be managed carefully
Coastal plains	Battery storage or flywheels	Easy access, stable temperatures, and proximity to infrastructure
Cold climates	Thermal storage or flywheels	Battery efficiency may drop, while thermal systems perform better
Tropical climates	Hybrid systems	Need to manage heat and humidity, combination of technologies is optimal

1.4.3 *Economic and Environmental Impacts*

When introducing ESSs, cost-effectiveness is crucial, especially when incorporating renewable energy sources. The cost–benefit balance, including energy savings, peak shaving, grid service revenue, and environmental externalities must be balanced with the capital and operating expenses.

1.4.3.1 Common Metrics Used to Evaluate the Cost-Effectiveness of Energy Storage Systems

1.4.3.1.1 Levelized Cost of Storage (LCOS)

It is among the most used criteria to assess how cost-effective energy storage is [34, 35, 39].

$$LCOS = \frac{Total \quad lifetime \quad cost}{Total \quad energy \quad over \quad lifetime} = \frac{C_{capex} + \sum_{1}^{N} \frac{C_{opex,t}}{(1+r)^{t}}}{E_{delivered}} \tag{1.28}$$

Additionally, it can be written as:

$$LCOS = \frac{C_{capex} + O\,\&\,M.\sum_{n=1}^{N} \frac{1}{(1+r)^{n}} - \frac{V_{res}}{(1+r)^{N+1}}}{c.DOD.RC\sum_{n=1}^{N} \frac{1-DEG.n}{(1+r)^{n}}} + \frac{P_{elec}}{\eta} \tag{1.29}$$

1.4.3.1.2 Net Present Value (NPV)

It calculates how profitable an energy storage investment will be. It is provided by [34, 35, 39]:

$$NPV = \sum_{n=1}^{N} \frac{R_{t} - C_{t}}{(1+r)^{n}} \tag{1.30}$$

1.4.3.1.3 Payback Period

It is the time needed to return the initial investment. It is expressed as [34, 35, 39]:

$$Payback \quad period = \frac{C_{initial}}{Annual \quad net \quad savings} \tag{1.31}$$

Also, batteries degrade over time. Hence, include cycle life and capacity in the model as:

$$C_{cost/cycle} = \frac{C_{capex} + C_{replacement}}{N_{cycles}} \tag{1.32}$$

1.4.4 Future Prospects and Innovations

1.4.4.1 Advancements in Energy Storage Technology

The need for scalability, efficiency, longevity, and cost-effectiveness is being met by advances in energy storage technology as global energy systems shift toward cleaner and more decentralized models. Important developments include performance enhancements and emerging storage technologies [1, 4, 8, 34, 35, 40].

1.4.4.2 Research and Development Directions

In order to lower costs, increase performance, and guarantee sustainability, the primary research and development directions are currently focused on materials innovation, which aims to create more affordable and abundant materials. The next developments involve hybrid storage technologies, enhanced system design, and strategies enabling easier recyclability and second-life-utilization. Last but not least, integrating smart grids with modern storage technologies and renewable energy sources is crucial [1, 4, 23–25, 40].

1.5 Conclusion

Future renewable energy systems will depend heavily on energy storage. This chapter has underlined the value of ESSs in mitigating the intermittency and variability inherent in renewable energy sources guaranteeing both technical dependability and financial feasibility for multiple purposes, such as enhancing grid stability, facilitating frequency management, controlling peak loads, and permitting a greater penetration of renewable energy sources.

References

1. Sahoo S, Timmann P (2023) Energy storage technologies for modern power systems: a detailed analysis of functionalities, potentials, and impacts. IEEE Access 11:49689–49729. https://doi.org/10.1109/ACCESS.2023.32745042
2. Rekioua D, Rekioua T, Elsanabary A, Mekhilef S (2023) Power management control of an autonomous photovoltaic/wind turbine/battery system. Energies 16:2286. https://doi.org/10.3390/en160522863
3. Rekioua D (2019) Hybrid renewable energy systems: optimization and power management control. Springer, Cham. https://doi.org/10.1007/978-3-030-34021-6
4. Rekioua D, Matagne E (2012) Optimization of photovoltaic power systems: modeling, simulation and control. Springer, Cham. https://doi.org/10.1007/978-1-4471-2403-05
5. Rekioua D, Mezzai N, Mokrani Z et al (2024) Effective optimal control of a wind turbine system with hybrid energy storage and hybrid MPPT approach. Sci Rep 14:30013. https://doi.org/10.1038/s41598-024-78847-9
6. Horzela-Miś A, Semrau J (2025) The role of renewable energy and storage technologies in sustainable development: simulation in the construction industry. Front Energy Res 13:1540423
7. Malik FH, Hussain GA, Alsmadi YMS, Haider ZM, Mansoor W, Lehtonen M (2025) Integrating energy storage technologies with renewable energy sources: a pathway toward sustainable power grids. Sustainability 17:4097. https://doi.org/10.3390/su17094097
8. Min C-G (2019) Analyzing the impact of variability and uncertainty on power system flexibility. Appl Sci 9:561. https://doi.org/10.3390/app9030561
9. Rekioua D (2024) Wind power electric systems: modeling, simulation and control. Springer, Cham. https://doi.org/10.1007/978-3-031-52883-5
10. Rekioua D, Rekioua T, Soufi Y (2015) Control of a grid-connected photovoltaic system. In: ICRERA, Palermo, Italy, pp 1382–1387. https://doi.org/10.1109/ICRERA.2015.7418634
11. Wu Y, Jiang D, Chen N, Zhang L, Qu L, Qian M (2018) Research on the impact of wind power generation with energy storage system on grid frequency stability. In: CCIS 2018, Nanjing, China, pp 1096–1100. https://doi.org/10.1109/CCIS.2018.8691186
12. Wu J, Zhang J, Dai Q, Yun L (2023) Research on influence and capacity configuration optimization of energy storage on power angle stability of large power grids. In: ICPES 2023, Chengdu, China, pp 240–245. https://doi.org/10.1109/ICPES59999.2023.10400127
13. Sun H, Zhao B, Xu S, Lan T, Li Z, Wu P (2025) Test models for stability/security studies of AC–DC hybrid power systems with high penetration of renewables. IEEE Trans Power Syst 40(1):957–969. https://doi.org/10.1109/TPWRS.2024.3408423
14. Wang D, Suo L (2025) Control-strategy characteristics of grid-forming/grid-following hybrid energy-storage systems for frequency support. In: NEESSC 2025, Hohhot, China, pp 162–166. https://doi.org/10.1109/NEESSC66038.2025.11199667

15. Wang D, Yan Y, Shen Y, Hu Z (2020) Active and reactive power coupling characteristics-based inertial and frequency control strategy of battery energy storage stations. In: APPEEC 2020, Nanjing, China, pp 1–5. https://doi.org/10.1109/APPEEC48164.2020.9220616
16. Kim M-J, Chae S-H, Moon Y-K (2020) Adaptive battery SOC estimation method for electric vehicle BMS. In: ISOCC 2020, Yeosu, Korea, pp 288–289. https://doi.org/10.1109/ISOCC50952.2020.9332950
17. Xiong R, Shen W (2019) Battery state of charge and state of energy estimation. In: Advanced battery management technologies for electric vehicles. Wiley, New York, pp 67–93. https://doi.org/10.1002/9781119481652.ch3
18. Sharma AK, Letha SS, Syal P, Rathee S, Kumar A (2025) State estimation and cell balancing for lithium-ion batteries powering electric vehicles. In: Smart electric and hybrid vehicles. Wiley, New York, pp 1–53. https://doi.org/10.1002/9781394225040.ch1
19. Song Y, Park M, Seo M, Kim SW (2019) Improved SOC estimation of lithium-ion batteries via a novel SOC–OCV curve estimation method using ECM. In: SpliTech 2019, Split, Croatia, pp 1–6. https://doi.org/10.23919/SpliTech.2019.8783149
20. Mohammadi M, Seward D, Fajri P, Sabzehgar R, Asrari A (2025) Machine learning-based modeling of EV power consumption. In: ITEC+EATS 2025, Anaheim, USA, pp 1–6. https://doi.org/10.1109/ITEC63604.2025.11097940
21. Abdelgabir H, Boynuegri AR, Elrayyah A, Sozer Y (2018) Small-signal modelling & stability analysis of nonlinear droop-controlled microgrids. In: APEC 2018, San Antonio, Texas, pp 3333–3339. https://doi.org/10.1109/APEC.2018.8341581
22. Chantola A, Sharma V, Singh D (2024) Centralized secondary control strategy of droop-controlled inverter-based microgrid. In: ICMICA 2024, Kurukshetra, India, pp 1–6. https://doi.org/10.1109/ICMICA61068.2024.10732485
23. Bevrani H (2009) Robust power system frequency control. Springer, Cham. https://doi.org/10.1007/978-0-387-84878-5_1
24. Yu L, Sun X, Li T, Zhong H, Li M, Shao Y (2022) Virtual inertia control and identification method of inertia parameters for doubly-fed units. In: SPIES 2022, Beijing, pp 958–963. https://doi.org/10.1109/SPIES55999.2022.100826642
25. Gundumalla VBK, Eswararao S (2018) Ramp rate control strategy for islanded DC microgrid with hybrid ESS. In: ICEES 2018, Chennai, India, pp 82–87. https://doi.org/10.1109/ICEES.2018.8442363
26. Engels J, Claessens B, Deconinck G (2020) Optimal combination of frequency control and peak shaving with battery storage systems. IEEE Trans Smart Grid 11(4):3270–3279. https://doi.org/10.1109/TSG.2019.2963098
27. Abbasi A, Khalid HA, Rehman H, Khan AU (2023) Dynamic load scheduling & peak shaving control in community HEMS. IEEE Access 11:32508–32522. https://doi.org/10.1109/ACCESS.2023.3255542
28. Garcesa A, Johnson NG, Nelson J (2025) Evaluating the stacked economic value of load shifting and microgrid control. Buildings 15:2378. https://doi.org/10.3390/buildings15132378
29. Ananda-Rao K, Ali R, Taniselass S, Baharudin NH (2016) Mic/rocontroller-based battery controller for peak shaving integrated with PV. In: CEAT 2016, Kuala Lumpur, 1–6. https://doi.org/10.1049/cp.2016.1262
30. Li Z, Zang C, Zeng P, Yu H, Li S (2018) Fully distributed hierarchical control of parallel grid-supporting inverters. IEEE Trans Industr Inform 14(2):679–690. https://doi.org/10.1109/TII.2017.2749424
31. Li YR, Nejabatkhah F, Tian H (2023) Energy management system (EMS) in smart hybrid microgrids. In: Smart hybrid AC/DC microgrids. IEEE, pp 155–183. https://doi.org/10.1002/9781119598411.ch6
32. Rekioua D, Roumila Z, Rekioua T (2008) Étude d'une centrale hybride photovoltaïque–éolien–diesel. J Renew Energ 11(4):623–633. https://doi.org/10.54966/jreen.v11i4.112
33. Shafiee Q, Naderi M, Bevrani H (2024) Interconnected microgrids: opportunities and challenges. In: Microgrids: dynamic modeling, stability and control. IEEE, New York, pp 263–284. https://doi.org/10.1002/9781119906230.ch7

34. Vatankhah Ghadim H, Haas J, Breyer C et al (2025) Clean technology cost projections for solar, wind, battery, and hydrogen. Sci Data 12:1670. https://doi.org/10.1038/s41597-025-05951-4
35. Schmidt O, Hawkes A, Gambhir A et al (2017) The future cost of electrical energy storage based on experience rates. Nat Energy 2:17110. https://doi.org/10.1038/nenergy.2017.110
36. Layedra J, Martínez M, Mercado P (2021) Levelized cost of storage for lithium batteries considering degradation and residual value. In: IEEE URUCON 2021. Montevideo, pp 127–131. https://doi.org/10.1109/URUCON53396.2021.9647345
37. Bose BK (2019) Grid energy storage systems. In: Power electronics in renewable energy systems and smart grid. IEEE, New York, pp 495–583. https://doi.org/10.1002/9781119515661.ch10
38. Ergun S, Dik A, Boukhanouf R, Omer S (2025) Large-scale renewable energy integration: technical challenges and storage innovations. Sustainability 17(3):1311. https://doi.org/10.3390/su1703131139
39. Wang L, Guo K, Nie C, Wu M (2024) Research on configuration of new energy storage capacity and cost evaluation for large energy bases. In: DCTS 2024, Zhuhai, 1–6. https://doi.org/10.1109/DCTS62535.2024.10939693
40. Yarga A, Daim TU, Alzahrani S, Burmaoglu S (2025) Technology assessment: energy storage technologies. In: Future-oriented technology assessment. IEEE, New York, pp 51–86. https://doi.org/10.1002/9781119909880.ch3

Chapter 2
Overview on Storage Systems Used in Renewable Energy System

2.1 Introduction

ESSs are essential for the large-scale deployment of RESs and for achieving sustainable energy systems. Consequently, it is now a strategic goal to incorporate advanced storage technologies into RESs [1–8]. This chapter presents various storage technologies and presents the criteria used to evaluate and select them [3–7] (Fig. 2.1). It highlights the essential performance indicators along with some applications. Through clear definitions, technical parameters, comparative tables, and relevant examples [9–11], this chapter gives readers a systematic understanding of how to combine storage solutions to renewable energy challenges.

2.2 Some Important Definitions

2.2.1 Introduction to ESSs

As explained in Chap. 1, since renewable energy sources are intermittent, ESSs are important for stabilizing power supply, balancing load, and improving grid reliability. There are various types of ESSs and an overview of the major types of ESSs, including mechanical, chemical, thermal, and electrical storage technologies are presented in this chapter [9–13]. Also, there are various factors influencing the storage selection based on some important parameters such as quick responses and seasonal storage [11–16].

D. Rekioua, *Energy Storage for Renewable Energy Systems*, Green Energy and Technology, https://doi.org/10.1007/978-3-032-19589-0_2

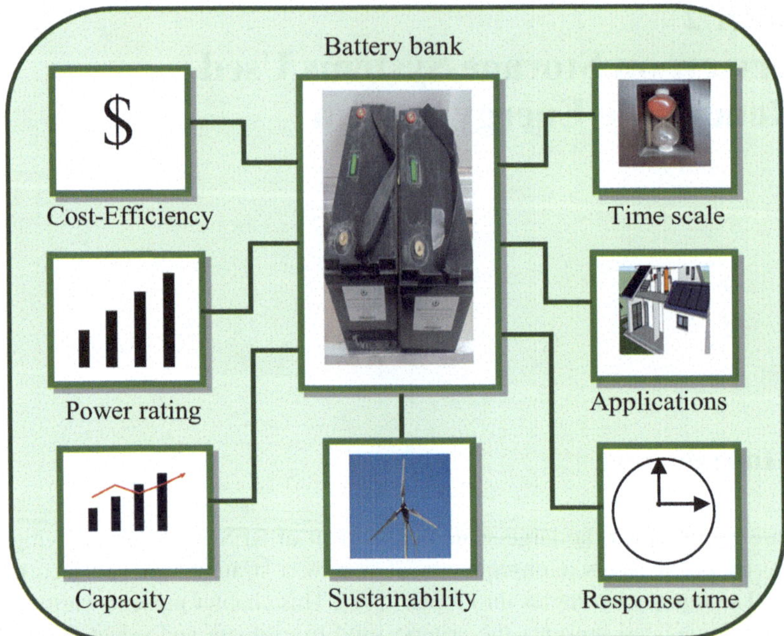

Fig. 2.1 Main factors influencing the selection of ESSs

2.2.2 Storage Capacity and Energy Density

They are the first factors to be considered in ESSs. Storage capacity is the energy stored in a system [9, 10, 14, 15]. Energy density indicates the quantity of energy stored per unit mass or volume. It describes the energy capacity of batteries, capacitors, or fuels relative to their weight (measured in Wh/kg) or size (measured in Wh/L).

$$Storage\ Capacity(Wh) = Storage\ \ capacity(Ah)\ \times\ Voltage(V) \quad (2.1)$$

The storage system's energy capacity (in kWh) can be expressed as follows:

$$E_{capacity} = P_{rated}\ t_{duration} \quad (2.2)$$

The storage capacity, energy density, applications, and operating periods of different energy storage systems are shown in Table 2.1 [9, 10, 12–15].

Gravimetric and volumetric energy densities are important in designing storage systems for different applications. The gravimetric energy density is the energy stored per kilogram of material in Wh/kg or J/kg. Choosing which energy density to prioritize depends entirely on different factors as physical space available, weight limitations, required autonomy or storage duration, …… [9, 10, 15, 16].

Table 2.1 Storage capacity, energy density, applications, and operational periods of different ES technologies

Storage technology	Storage capacity	Applications	Operational periods
Lead-acid batteries	kW to MW	Backup power, Small-scale renewable energy integration	**ST**
Lithium-ion batteries	kW to GW	Electric vehicles, Grid stabilization, Renewable integration	**ST to MT**
Nickel-cadmium batteries	kW to MW	Industrial applications, Critical backup	**ST**
Sodium-sulfur batteries	MW	Grid-scale storage, Renewable integration	**MT**
Flow batteries	kW to MW	Large-scale renewable integration, Grid services	**MT**
Supercapacitors	kW to MW	Power quality, short-term grid balancing, Quick charge/discharge applications	**VST**
Flywheels	kW to MW	Frequency regulation, Uninterruptible power supply, Grid services	**VST**
Pumped hydro storage	GW	Operate at the scale of the electrical grid**,** Renewable energy management	**LT**
Compressed air energy storage (CAES)	MW to GW	Large-scale energy storage, Reducing electricity demand during peak hours	**LT**
Thermal energy storage	MW to GW	Concentrated solar power, District heating	**LT**
Hydrogen storage	kW to GW	Seasonal storage, Renewable integration, transportation	**LT**

$$Gravimetric \quad energy \quad density = \frac{Total \quad energy \quad stored \quad (Wh)}{Mass \quad (kg)} \tag{2.3}$$

While the volumetric energy density (energy per volume) is the energy stored per liter or cubic meter of volume in Wh/L or J/m^3.

$$Volumetric \quad energy \quad density = \frac{Total \quad energy \quad stored \quad (Wh)}{Volume \quad (kg\, or\, m^3)} \tag{2.4}$$

2.2.3 *Large-Scale Storage*

This term describes energy storage systems that are designed to manage and store massive amounts of energy, usually in the megawatt (MW) to gigawatt (GW) range [9–15, 17–22]. These systems are primarily used for grid applications to control supply and demand, stabilize the electrical grid, integrate large renewable energy resources, and provide backup power during power interruptions (Fig. 2.2).

The most commonly used storage technologies are illustrated (Fig. 2.3) [23–25].

2.2.4 *Quick Response Applications*

It means ESSs and technologies that are designed to react and deliver power almost instantly in reaction to variations in grid demand, power quality issues, or fluctuations in energy generation. These applications have importance for preserving grid stability and ensuring power quality, and supporting systems where rapid energy delivery is essential. The main features of rapid response applications are the ability to quickly store or release energy within seconds or milliseconds, obtaining with high output power, suitable for frequency regulation, voltage support, and mitigating sudden load changes (Fig. 2.4).

The common technologies for quick response applications include [28–33]:

- Flywheels: are frequently utilized for grid frequency management and offer quick energy discharge.
- Lithium-ion batteries: can be configured for high-power applications with quick response, such as grid stabilization.
- Supercapacitors: with an ability to provide rapid response during power interruptions.

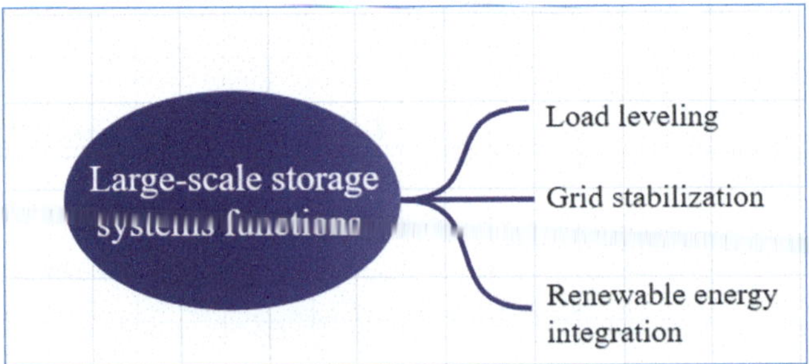

Fig. 2.2 Main large-scale storage systems functions

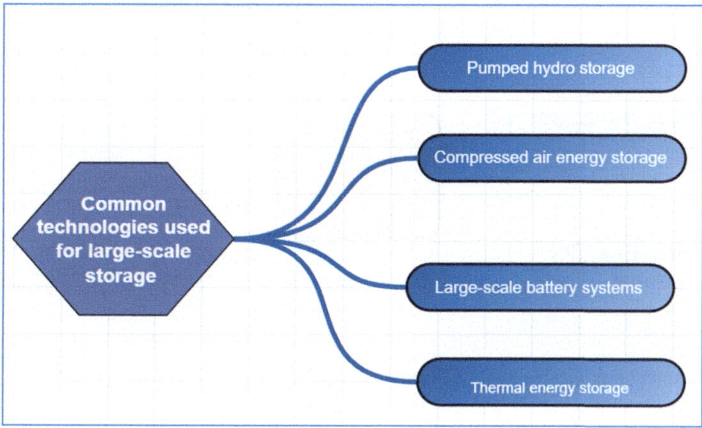

Fig. 2.3 Common technologies used for large scale storage

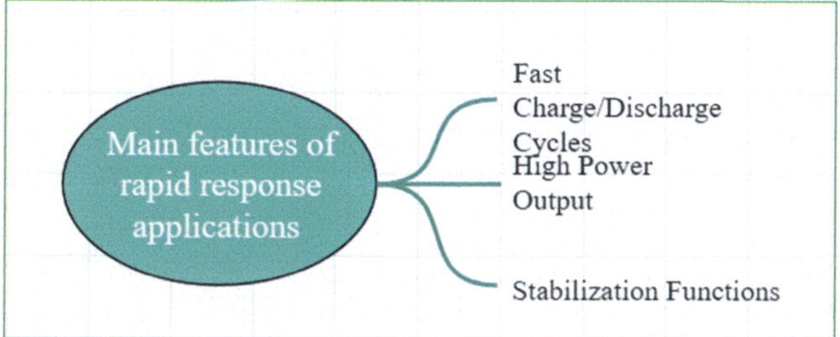

Fig. 2.4 Main features of rapid response applications

2.2.5 Seasonal Storage

It refers to energy storage systems designed to store energy over long periods, typically from one season to another [3, 7]. Seasonal storage is crucial for managing the seasonal variability of renewable energy resources [19, 20]. The main technologies used for seasonal storage are shown in Fig. 2.5 [17–22].

2.2.6 Power Rating and Response Time

Power rating is the maximum power the storage system delivered at a given time, while response time refers to how quickly the system can start discharging or charging energy [9, 10].

The power rating equation is:

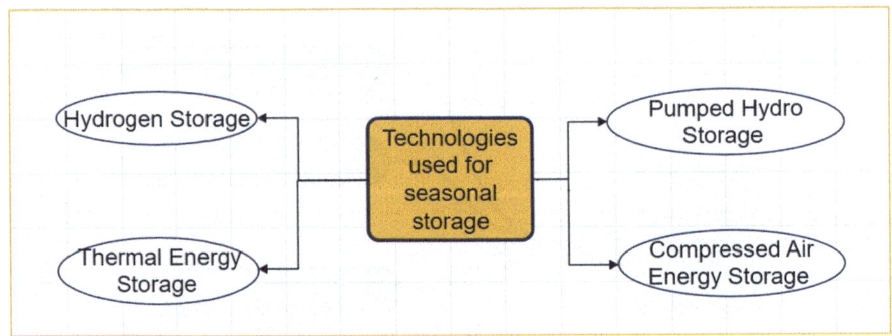

Fig. 2.5 Main technologies used for seasonal storage

$$P = \frac{E}{t_{discharge}} \tag{2.5}$$

The response time can be written as:

$$t_a = \frac{V_{ref} - V_{actual}}{dV / dt} \tag{2.6}$$

2.2.7 *Efficiency*

Efficiency is a key performance indicator for ESSs [9, 10, 13, 15].

$$\eta_{sto} = \frac{E_{out}}{E_{in}} \tag{2.7}$$

2.2.8 *Cost and Economic Feasibility*

Costs associated with installation, maintenance, and lifetime are important in the selection of storage technologies. Indeed, cost and economic feasibility are essential for selecting appropriate ESS technologies. By analyzing LCOS, payback period, and NPV, decision-makers can assess whether a system is financially viable for grid support, renewable integration, or standalone applications. LCOS is among the most common metrics for evaluating the cost-effectiveness of ESS [14–18].

$$LCOS = \frac{C_{capex} + \sum_{n=1}^{N} \frac{C_{opex,t}}{(1+r)^n}}{E_{delivered}} \tag{2.8}$$

Where: C_{capex} is the Initial investment or Capital, $C_{opex,t}$ is the operational & maintenance costs that includes regular maintenance, replacements, ….., and $E_{delivered}$ is the energy delivered:

$$C_{capex} = C_{storage} + C_{inverter} + C_{installation} + C_{infrastructure} \tag{2.9}$$

$$C_{opex} = C_{maintenance} + C_{replacement} + C_{minitoring} \tag{2.10}$$

Example

If a Li-ion battery system costs \$150/kWh for 1000 kWh, where the C_{capex} = \$150,000, the C_{opex} = \$3000/year × 10 years = \$30,000 and

$$E_{delivered} = 1000kWh \times 365 \times 10 years \times 0.9(efficiency) = 3285000kWh$$

To calculate the LCOS, we have applied the Eq. (2.8):

$$LCOS = \frac{C_{capex} + \sum_{n=1}^{N} \frac{C_{opex,t}}{(1+r)^n}}{E_{delivered}} = \frac{150 + 3.00 \times 10}{1000 \times 365 \times 10 \times 0.9} = 0.05479 \approx 0.055\$ / kWh$$

An energy storage investment's financial viability is assessed by its Net Present Value (NPV). It is given by:

$$NPV = \sum_{t=0}^{n} \frac{C_t}{(1+r)^n} \tag{2.11}$$

Example

For NPV calculation for a battery storage project, we have

- Initial investment cost (year 0) = 50,000 \$
- Annual net savings from energy arbitrage and peak shaving = 12,000 \$
- Project lifetime = 5 years
- Discount rate = 8%

$$NPV = -50 + \frac{12.00}{(1+0.08)^1} + \frac{12.00}{(1+0.08)^2} + \frac{12.00}{(1+0.08)^3} + \frac{12.00}{(1+0.08)^4} + \frac{12.00}{(1+0.08)^5}$$

$$NPV = -50 + 11.11 + 10.291 + 9.528 + 8.818 + 8.166 + 8.17 = -2.086\$$$

$$NPV < 0$$

It is negative, so this project would not be considered economically feasible under these assumptions. However, if savings increased (or cost decreased), NPV could become positive.

The payback period is the time needed to return to the initial investment. It is more attractive to have a shorter payback time. It is expressed as follows:

$$Payback \quad Period = \frac{C_{initial}}{Annual \quad net \quad savings} \tag{2.12}$$

Example

The total battery system cost is: \$10,000 and the annual savings from diesel energy consumption is about \$1500. Calculate the Payback period.

$$Payback \quad Period = \frac{10.000}{1.500} = 6.66 \approx 6.7\,years$$

Further assessment Internal Rate of Return (IRR), Benefit-to-Cost Ratio (BCR), Total Cost of Ownership (TCO), and other metrics are also useful.

The TCO is the sum of all costs across the lifecycle.

$$TCO = C_{capex} + \sum_{t=1}^{n} \frac{C_{opex,t}}{(1+r)^t} \tag{2.13}$$

It can be also written as:

$$TCO = C_{initial} + C_{O\&M} + C_{replacement} - C_{residual} \tag{2.14}$$

Example

Calculate the TCO of a Li-ion battery system of 50 kWh with the different conditions:

- Initial cost ($C_{initial} = 20{,}000\$$)
- O&M costs over 10 years ($C_{O\&M} = 2000\$$)
- Replacement after year 7 = 6000\$
- Residual value at year 10 = 1000\$

We use Eq. 2.14:

$$TCO = 20.000 + 2.000 + 6.000 - 1.000 = 27.000\$$$

The BCR is the ratio of total benefits to total costs.

$$BCR = \frac{\sum_{t=1}^{N} \frac{B_t}{(1+r)^t}}{\sum_{t=1}^{N} \frac{C_t}{(1+r)^t}} \tag{2.15}$$

It can also be written as:

$$BCR = \frac{Total \quad benefits \quad over \quad lifetime}{TCO} \tag{2.16}$$

Example

We consider a system which saves \$3000/year in energy bills with a lifetime of 10 years. Calculate the total benefits and deduce the BCR.

The total benefit is: 10 × 3000 = 30,000$

$$BCR = \frac{30.000}{30.000 - 3.000} = \frac{30.000}{27.000} = 1.11$$

$BCR > 1$ thus the investment is economically viable.

We can compare three indicators for some storage technology (Table 2.2).

2.2.9 Durability and Life Cycle

The longevity and cycle life of different storage systems vary, with some technologies degrading faster than others over repeated use. Durability is the ability of an ESS to continue operating over time, especially its capacity to charge and discharge

Table 2.2 Storage examples in terms of C_{capex}, C_{opex} and $LCOS$

Storage type	C_{CAPEX}	C_{OPEX}	LCOS
Lithium-ion battery	++++	+++	+
Pumped hydro	+++	+	–
Flow battery	++	++	++
CAES	+++	++	++
Supercapacitors	++	+	++++

With: ++++ Very high, +++ High, ++ Moderate, + Low –Very low

efficiently without suffering appreciable deterioration [37–39]. Usually, the life cycle indicates how many cycles of charging and discharging the system is capable of before its capacity drops below a practical level [14, 15].

2.3 Different Energy Storage Technologies

Some works provide an overview of various ESSs and how well they work with RESs. Electrochemical energy storage systems (ESS) as batteries, fuel cells, and flow batteries. Mechanical storage methods include flywheel energy storage (FES), compressed air energy storage (CAES), and pumped hydroelectric energy storage (PHES). In CAES, compressed air is kept in caverns and then released to generate electricity when demand is high. Flywheel energy storage (FES) allows kinetic energy to be stored in a rotating flywheel. Many factors, such as cost, efficiency, environmental impact, and technical availability, affect the decision to use a mechanical ESS.

Superconducting magnetic energy storage (SMES) devices store energy in a superconducting coil, which transmits electricity without resistance. Fast reaction times and almost instantaneous discharge capabilities are advantages of SMES systems. However, superconducting materials are expensive due to their high cost and the need for specialized cooling equipment. Because supercapacitors store energy in electrostatic fields and generate high power output with fast response times, they are used for applications like grid stability and regenerative braking in electric vehicles. Thermal energy storage (TES), which uses a range of materials to store energy as heat, is ideal for long-duration storage even though it typically generates less power than other ESS technologies. The following (Fig. 2.6) are the four main types of ESSs [4, 15–20].

A number of criteria as system requirements, prices influence the choice of an ESS for PV and wind systems. Since different technologies have distinct performance characteristics and are better suited to particular use cases, there is not a standard approach for selecting an energy storage system for PV or wind applications.

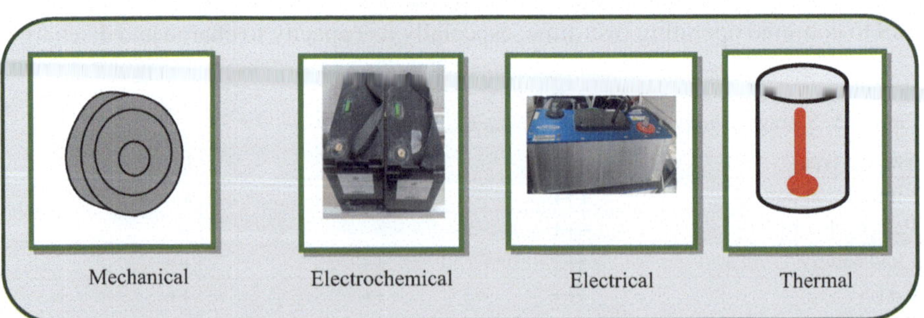

Fig. 2.6 Primary categories of ESSs

2.3.1 Electrochemical Storage

Numerous electrochemical storage (ESS) technologies, as fuel cells and batteries, are used. ESSs offers a number of benefits, it is modular and adaptable and can be flexible to meet a range of power needs. Numerous electrochemical ESSs offer excellent round-trip efficiency rates, making it extremely efficient as well. ES does have certain drawbacks [16] (Fig. 2.7).

2.3.2 Batteries Energy Storage Systems (BESSs)

Energy storage devices use a wide variety of battery types. Flow batteries are commonly used in combination with PV and wind systems for energy storage. The three types of flow batteries that are vanadium flow batteries (VFB), zinc-bromine flow batteries (ZBFB), and hybrid flow batteries.

2.3.3 Hydrogen Energy Storage (HES)

HES provides a range of advantages in the context of energy storage. It serves as an effective energy storage solution for remote or off-grid regions since it can be produced from RESs and stored for a long time. Additionally, it is a flexible and safe energy source. Nevertheless, there are certain drawbacks to HES. One issue is that hydrogen production and storage can be more expensive than alternative energy

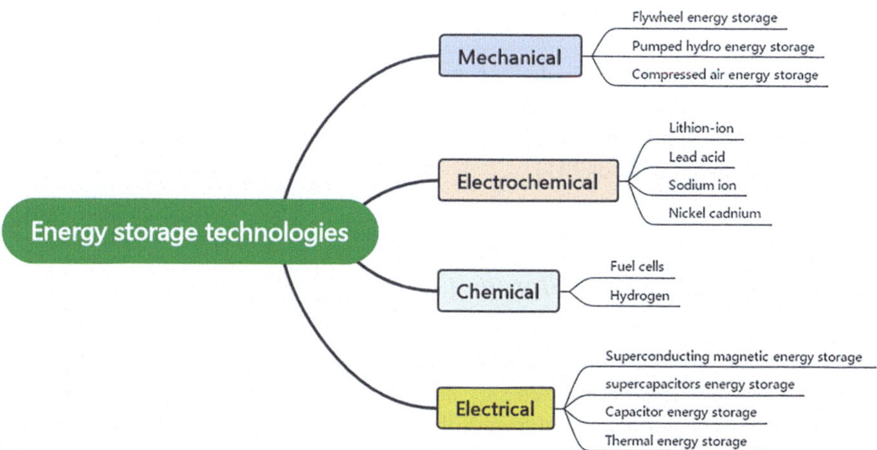

Fig. 2.7 Energy storage technologies

storage methods. Additionally, HES necessitates specialized infrastructure, including as fuel cells and pipelines, which can be costly to construct and maintain [16–27].

2.3.4 Mechanical Storage

2.3.4.1 Pumped Hydro Energy Storage (PHES)

Two reservoirs at various elevations are used in pumped hydro energy storage (PHES) to produce and store energy. Water is pumped from the lower reservoir to the higher reservoir when the grid has extra power, to store the etra energy as gravitational potential energy. PHES systems are an excellent option, especially for long-term storage [25].

2.3.4.2 Compressed Air Energy Storage (CAES)

Energy is produced by compressing air, which is then stored in subsurface tanks or caverns.To generate power, the compressed air is released and allowed to expand through a turbine. Unlike other systems for energy storage, CAES installations require access to suitable geological formations and can be more expensive to construct [22].

2.3.4.3 Flywheel Energy Storage

The flywheel rotates rapidly inside a vacuum-sealed container to reduce energy waste and friction. When power is available, the flywheel accelerates to store energy, and when electricity is needed, a generator converts the flywheel's rotational energy back into energy [23, 28–30].

2.3.4.4 Gravity Energy Storage (GES)

By raising weights to a specific height, it stores potential energy. If the weights are released, they fall and power a generator, releasing the stored energy for later use. Water is pumped to a higher reservoir, where it can be released to generate energy during high-demand periods [31].

2.3.5 Thermal Energy Storage (TES)

Power generation, industrial operations, and building heating and cooling are a few of its uses. Storage media,, molten salt, and phase transition materials, are used by TES systems. When energy is required, the stored heat is released to provide heat, power, or both [32, 33].

2.3.6 *Electrical Energy Storage*

2.3.6.1 Supercapacitor Energy Storage (SES)

It relies on electrostatic fields for energy storage. SES devices may store and release more energy faster due to their significantly increased energy density [34, 35].

2.3.6.2 Superconducting Magnetic Energy Storage (SMES)

SMES systems typically consist of superconducting coils, power conditioning units, and cryogenic cooling systems. When a current flows through a coil, a magnetic field is produced that can be used to store energy [4, 36, 37].

2.3.6.3 Capacitor Energy Storage (CES)

Two capacitors are separated by an insulating material called a di-electric and two conducting surfaces. One plate of the capacitor gets negatively charged and positively charged, respectively when a voltage is applied, creating an electric field between the plates [38].

2.3.7 *State of Charge (SOC)*

The SOC indicates how much energy is remaining in a battery in relation to its maximum capacity; in other words, it indicates how "full" the battery is at any particular time [39, 40].

The mathematical definition is:

$$SOC = \frac{Q_{remaining}}{Q_{nominal}}.100 \tag{2.17}$$

The battery status for each *SOC* value is explained in Table 2.3, where 100% represents the maximum stored energy available, [70–90%] is recommended for extended life, [20–50%] is acceptable during operation, and [0–20%] should be avoided because it increases degradation.

SOC must always remain within defined limits in order to ensure safe operation:

$$SOC_{\min}\ ''\ SOC''\ SOC_{miax} \tag{2.18}$$

Table 2.3 State of charge evolution

SOC (%)	Battery Status
100%	Fully charged
70–90%	Optimal operating range
20–50%	Partially discharged
0–20%	Deep discharge

2.3.8 *Depth of Discharge* (DOD)

It is a crucial measure that shows how much of a battery's total capacity has been utilized during a discharge cycle [40–44].

$$DOD = \frac{Energy \quad discharged}{Total \quad battery \quad capacity}.100 \tag{2.19}$$

The relation between *SOC* and *DOD* is:

$$SOC + DOD = 100\% \tag{2.20}$$

Figure 2.8 effectively and clearly illustrates the relationship between the *SOC* and the *DOD* in a battery system, which together represent the battery's current energy status.

The notion of SOC_{max} and SOC_{min}, which specify the recommended operating limitations of the battery to maintain its health and ensure long-term performance, is also highlighted in the diagram.

Example
It is considered in this example, a 100 Ah battery which delivers [40, 55, 60] Ah before being recharged. Table 2.4 shows how to calculate the DOD for a 100 Ah battery delivering different amounts of energy before being recharged.

2.3.9 *Battery Lifetime*

The estimated cycle life N_{cycle} of the battery is calculated as [40–42]:

$$N_{cycle} = N_0 \cdot \left(\frac{DOD_{ref}}{DOD_{actual}} \right)^k \tag{2.21}$$

Then, the lifetime increases to N_1 cycles:

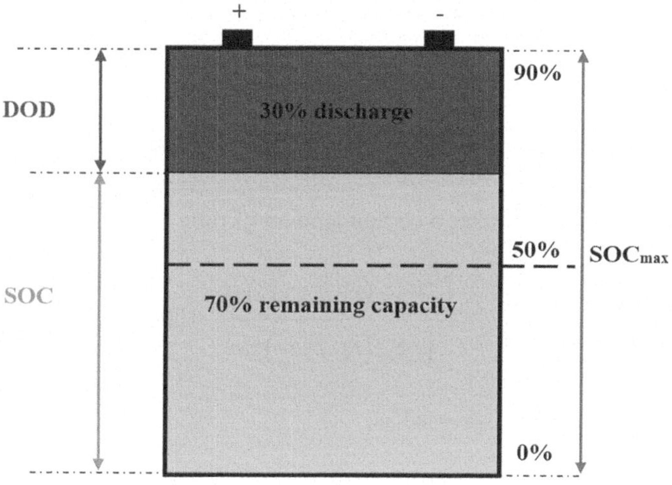

Fig. 2.8 Relationship between *SOC* and *DOD*

Table 2.4 *DOD* Calculation for a 100 Ah Battery

Delivered Capacity (Ah)	Nominal Capacity (Ah)	*DOD* (%)
40 ah	100 ah	40%
55 ah	100 ah	55%
60 ah	100 ah	60%

$$\% = \frac{N_1 - N_0}{N_0}.100 \tag{2.22}$$

Example

If we have the following values: N_0 = 2000 cycles, DOD_{ref} = 100%, DOD_{actual} = [60%–100%], and k = 1.1, we can find the different results for various values of DOD_{actual} given in Table 2.5.

2.3.10 Efficiency of ESSs

Battery charging efficiency measures the rate at which a battery transforms electrical energy input into chemical energy stored during charge [40–42].

$$\eta_{ch} = \frac{Battery \quad stored \quad energy}{Energy \quad supplied \quad during \quad charging}.100 = \frac{E_{sto}}{E_{in}}.100 \tag{2.23}$$

Table 2.5 Obtained results based on variation of DOD_{ref}

DOD_{actual}	N_c (cycles)	% improvement Lifetime
100%	2000	0%
70%	2850	+42.5%
60%	3450	+72.5%

If the process of charging takes a certain amount of time t_{ch}:

$$\eta_{ch} = \frac{\int_0^{t_{ch}} V_{disch}(t) \,.\, I_{disch}(t).dt}{\int_0^{t_{ch}} V_{ch}(t) \,.\, I_{ch}(t).dt}.100 \tag{2.24}$$

Discharging efficiency is expressed as:

$$\eta_{disc} = \frac{E_{out}}{E_{sto}}.100 \tag{2.25}$$

Charge and discharge efficiencies are taken into account by the round-trip efficiency:

$$\eta_{rt} = \eta_{ch} \cdot \eta_{disc} \tag{2.26}$$

Battery longevity and energy efficiency are increased when intelligent control systems are combined with renewable energy.

Example
Here is an application of a lithium-ion battery with its various parameters: Nominal capacity: C_{nom} = 100 Ah; nominal voltage: V_{nom} = 48 V. The usable energy, if full is:

$$E_{nom} = V_{nom}.C_{nom} = 48 \times 100 = 4800Wh$$

One charge event and one discharge event (often between SOC 20% and SOC 90%; not full 0–100% cycles) constitute operation.

In this example, the following energy quantities are assumed during a single cycle:

- Energy supplied during charging at the charger/inverter input: E_{in} = 5000 Wh (with losses from cables and converters)
- Saved battery energy: E_{saved} = 4600 Wh.
- The energy supplied to the load during discharge, after converter losses, is equal to E_{out} = 4300 Wh.

The charging efficiency is:

$$\eta_{ch} = \frac{4600}{5000} = 0.92 = 92\%$$

The discharging efficiency is:

$$\eta_{dish} = \frac{4300}{4600} = 0.9348 = 93.48\%$$

And then the round trip efficiency:

$$\eta_{rt} = \frac{4300}{5000} = 0.86$$

Alternatively:

$$\eta_{rt} = 0.92 \times 0.9348 = 0.86 = 86\%$$

2.4 Conclusion

The importance of ESSs in RESs has been discussed in this chapter. By comparing a wide range of technologies, from lithium-ion batteries and supercapacitors to pumped hydro and hydrogen storage, we have shown how technical requirements, economic viability, and application context influencing the choice of the optimum energy storage system [43–48]. There isn't a solution that works for every system, and the best storage option is determined by a number of variables, including cost, location, response time, power consumption, and environmental constraints.

References

1. Palanisamy S, Shanmugasundaram L, Chenniappan S (2023) Energy storage systems for smart power systems. In: Artificial intelligence-based smart power systems. IEEE, New York, pp 99–114. https://doi.org/10.1002/9781119893998.ch5
2. Rekioua D (2020) Hybrid renewable energy systems overview. In: Hybrid renewable energy systems. Green energy and technology. Springer, Cham. https://doi.org/10.1007/978-3-030-34021-6_1
3. Rekioua D, Rekioua T, Elsanabary A, Mekhilef S (2023) Power management control of an autonomous photovoltaic/wind turbine/battery system. Energies 16(5):2286. https://doi.org/10.3390/en16052286
4. Cui L, Li D, Wen J, Jiang Z, Cheng S (2008) Application of superconducting magnetic energy storage unit to damp power system low frequency oscillations. In: 2008 third international conference on electric utility deregulation and restructuring and power technologies. IEEE, New York, pp 2251–2257. https://doi.org/10.1109/DRPT.2008.4523785
5. Rekioua D (2019) Hybrid renewable energy systems: optimization and power management control. Springer, Cham. https://doi.org/10.1007/978-3-030-34021-6

6. Rekioua D, Matagne E (2012) Optimization of photovoltaic power systems: modeling, simulation and control. Springer. https://doi.org/10.1007/978-1-4471-2403-0
7. Rekioua D (2020) Storage in hybrid renewable energy systems. In: Hybrid renewable energy systems. Springer, Cham. https://doi.org/10.1007/978-3-030-34021-6_4
8. Ould Amrouche S, Rekioua D, Rekioua T, Bacha S (2016) Overview of energy storage in renewable energy systems. Int J Hydrog Energy 41(45):20914–20927. https://doi.org/10.1016/j.ijhydene.2016.06.243
9. Li YR, Nejabatkhah F, Tian H (2023) Renewable energy, energy storage, and smart interfacing power converters. In: Smart hybrid AC/DC microgrids: power management, energy management, and power quality control. IEEE, New York, pp 21–54. https://doi.org/10.1002/9781119598411.ch2
10. Schischke E, Grevé A, Ehrenstein U, Doetsch C (2024) Overview of energy storage technologies besides batteries. In: Passerini S, Barelli L, Baumann M, Peters J, Weil M (eds) Emerging battery technologies to boost the clean energy transition. Springer, Cham. https://doi.org/10.1007/978-3-031-48359-2_4
11. Nkembi AA, Simonazzi M, Santoro D, Cova P, Delmonte N (2024) Comprehensive review of energy storage systems characteristics and models for automotive applications. Batteries 10(3):88. https://doi.org/10.3390/batteries10030088
12. Akinyele D, Rayudu R (2014) Review of energy storage technologies for sustainable power networks. Sustain Energy Technol Assess 8:74–91. https://doi.org/10.1016/j.seta.2014.07.004
13. Manjulata B, Saurav R, Sheila M, Swetha GS (2023) Introduction to smart energy systems in recent trends. In: Applications of big data and artificial intelligence in smart energy systems, vol 1. River Publishers, pp 1–24
14. Talebi S, Aly HH (2024) A review of battery energy storage system optimization: current state-of-the-art and future trends. In: 2024 international conference on green energy, computing and sustainable technology (GECOST). IEEE, New York, pp 116–120. https://doi.org/10.1109/GECOST60902.2024.10474701
15. Schmidt O, Melchior S, Hawkes A, Staffell I (2019) Projecting the future levelized cost of electricity storage technologies. Joule 3(1):81–100. https://doi.org/10.1016/j.joule.2018.12.008
16. Migliari L, Cocco D, Petrollese M (2025) Levelized cost of storage (LCOS) of battery energy storage systems (BESS) deployed for photovoltaic curtailment. In: Rekioua D, Matagne E (eds) Modeling of storage systems. Optimization of photovoltaic power systems. Springer, Cham. https://doi.org/10.1007/978-1-4471-2403-0_5
17. Aghmadi A, Mohammed OA (2024) Energy storage systems: technologies and high-power applications. Batteries 10(4):141. https://doi.org/10.3390/batteries10040141
18. Clapp AL (2006) Storage batteries. In: NESC handbook: a discussion of the National Electrical Safety Code. IEEE, New York, pp 165–169. https://doi.org/10.1002/9781118098974.ch9
19. Rekioua D (2023) Energy storage Systems for Photovoltaic and Wind Systems: a review. Energies 16(9):3893. https://doi.org/10.3390/en16093893
20. Nazaralizadeh S, Banerjee P, Srivastava AK, Famouri P (2024) Battery energy storage systems: a review of energy management systems and health metrics. Energies 17(5):1250. https://doi.org/10.3390/en17051250
21. Zaouche F, Rekioua D, Gaubert J-P, Mokrani Z (2017) Supervision and control strategy for photovoltaic generators with battery storage. Int J Hydrog Energy 42(30):19536–19555. https://doi.org/10.1016/j.ijhydene.2017.06.107
22. Olabi AG, Allam MA, Abdelkareem MA, Deepa TD, Alami AH, Abbas Q, Alkhalidi A, Sayed ET (2023) Redox flow batteries: recent development in main components, emerging technologies, diagnostic techniques, large-scale applications, and challenges and barriers. Batteries 9(8):409. https://doi.org/10.3390/batteries9080409
23. Dicks AL, Rand DAJ (2018) Introducing fuel cells. In: Fuel cell systems explained. https://doi.org/10.1002/9781118706992.ch1
24. Pukrushpan JT, Stefanopoulou AG, Peng H (2004) Fuel cell system model: auxiliary components. In: Control of fuel cell power systems. Springer, Cham, pp 23–48. https://doi.org/10.1007/978-1-4471-3792-4_2

25. Bensmail S, Rekioua D, Azzi H (2015) Study of hybrid photovoltaic/fuel cell system for stand-alone applications. Int J Hydrog Energy 40(39):13820–13826. https://doi.org/10.1016/j.ijhydene.2015.04.013
26. Mezzai N, Rekioua D, Rekioua T, Mohammedi A, Idjdarane K, Bacha S (2014) Modeling of hybrid photovoltaic/wind/fuel cells power system. Int J Hydrog Energy 39(27):15158–15168. https://doi.org/10.1016/j.ijhydene.2014.06.015
27. Rekioua D, Bensmail S, Bettar N (2014) Development of hybrid photovoltaic-fuel cell system for stand-alone application. Int J Hydrog Energy 39(3):1604–1611. https://doi.org/10.1016/j.ijhydene.2013.03.040
28. Omsin P, Sharkh SM, Moshrefi-Torbati M (2019) A hybrid SS-CAES system with a battery. In: 2019 16th international conference on electrical engineering/electronics, computer, telecommunications and information technology (ECTI-CON). IEEE, New York, pp 159–162. https://doi.org/10.1109/ECTI-CON47248.2019.8955397
29. Tammam B, Adrian I, Rafic Y, Jean P (2012) A new multi-hybrid power system for grid-disconnected areas wind—diesel—compressed air energy storage. In: 2012 international conference on renewable energies for developing countries (REDEC). IEEE, New York, pp 1–8. https://doi.org/10.1109/REDEC.2012.6416683
30. Amiryar ME, Pullen KR (2017) A review of flywheel energy storage system technologies and their applications. Appl Sci 7(3):286. https://doi.org/10.3390/app7030286
31. Aissou R, Rekioua T, Rekioua D, Tounzi A (2016) Application of nonlinear predictive control for charging the battery using wind energy with permanent magnet synchronous generator. Int J Hydrog Energy 41(45):20964–20973. https://doi.org/10.1016/j.ijhydene.2016.05.249
32. Rekioua D, Roumila Z, Rekioua T (2008) Étude d'une centrale hybride photovoltaïque–éolien–diesel. J Renew Energ 11(4):623–633. https://doi.org/10.54966/jreen.v11i4.112
33. Yu C, Wang Z, Zhao W, Han B, Xu M, Liu Z (2022) Research on double closed-loop control method of wind power gravity energy storage system. In: 2022 5th international conference on power and energy applications (ICPEA). IEEE, New York, pp 704–708. https://doi.org/10.1109/ICPEA56363.2022.10052486
34. Kwasi-Effaha CC, Okpako O (2025) Comprehensive review of emerging trends in thermal energy storage mechanisms, materials and applications. Front Energy Res 13:1651471. https://doi.org/10.3389/fenrg.2025.1651471
35. Kale P, Gondane P, Bhutada P, Patil A, Patil Y, Yadav R (2025) Phase change materials in thermal energy storage: a comprehensive review of properties, advances, and challenges. In: 2025 international conference on sustainable energy technologies and computational intelligence (SETCOM). IEEE, New York, pp 1–6. https://doi.org/10.1109/SETCOM64758.2025.10932354
36. Burke AF (2007) Comparisons of lithium-ion batteries and ultracapacitors in hybrid-electric vehicle applications. In: EET-2007 European Ele-drive conference, Brussels, Belgium
37. Rekioua D, Kakouche K, Babqi A, Mokrani Z, Oubelaid A, Rekioua T, Azil A, Ali E, Alaboudy AHK, Abdelwahab SAM (2023) Optimized power management approach for photovoltaic systems with hybrid battery-supercapacitor storage. Sustainability 15(19):14066. https://doi.org/10.3390/su151914066
38. Tam K-S, Yarali A (1993) Operation principle and applications of multiterminal superconductive magnetic energy storage systems. IEEE Trans Energy Convers 8(1):54–62. https://doi.org/10.1109/60.207406
39. Li T, Hu W, Zhang B, Zhang G, Zhang Z, Chen Z (2021) SMES damping controller design and real-time parameters tuning for low-frequency oscillation. IEEE Trans Appl Supercond 31(8):3800604. https://doi.org/10.1109/TASC.2021.3099759
40. Strajnikov P, Peretz MM, Kuperman A (2022) Reduced-rating-electronic-capacitor-based DC links for power electronics supported distributed energy resources. IEEE Trans Smart Grid 13(5):3943–3953. https://doi.org/10.1109/TSG.2021.3120773
41. Wei Z, Sun X, Li Y, Liu W, Liu C, Lu H (2025) A joint estimation method for the SOC and SOH of lithium-ion batteries based on AR-ECM and data-driven model fusion. Electronics 14(7):1290. https://doi.org/10.3390/electronics14071290

42. Nishanth G, Krishnan MM, Parandhaman B et al (2025) Hardware implementation of EKF based SOC estimate for lithium-ion batteries in electric vehicle applications. Sci Rep 15:15551. https://doi.org/10.1038/s41598-025-99931-8
43. Madani SS, Shabeer Y, Allard F, Fowler M, Ziebert C, Wang Z, Panchal S, Chaoui H, Mekhilef S, Dou SX, See K, Khalilpour K (2025) A comprehensive review on lithium-ion battery lifetime prediction and aging mechanism analysis. Batteries 11(4):127. https://doi.org/10.3390/batteries11040127
44. Prashanth GK, Akolkar HN, Haghi AK, Rao S (2025) Fundamentals of lithium-ion batteries. In: New generation of Lithium-ion batteries for renewable energy storage. River Publishers, pp 1–34
45. Belaid S, Rekioua D, Oubelaid A, Ziane D, Rekioua T (2022) Proposed hybrid power optimization for wind turbine/battery system. Period Polytech Electr Eng Comput Sci 66(1):60–71. https://doi.org/10.3311/PPee.18758
46. Smith K, Baggu M, Friedl A, Bialek T, Schimpe MR (2017) Performance and health test procedure for grid energy storage systems. In: 2017 IEEE Power & Energy Society General Meeting. IEEE, New York, pp 1–5. https://doi.org/10.1109/PESGM.2017.8274326
47. Rekioua D, Mezzai N, Mokrani Z et al (2024) Effective optimal control of a wind turbine system with hybrid energy storage and hybrid MPPT approach. Sci Rep 14:30013. https://doi.org/10.1038/s41598-024-78847-9
48. Diao H et al (2025) Research progress and prospect of salt cavern flow battery for large scale energy storage. In: 2025 6th international conference on clean energy and electric power engineering (ICCEPE). IEEE, New York, pp 513–519. https://doi.org/10.1109/ICCEPE66357.2025.11193218

Chapter 3
Different Configurations of Storage in RESs

3.1 Introduction

The integration of storage architectures (single, dual, hybrid, and multi-storage) with wind turbines and photovoltaic (PV) systems to convert variable generation into reliable electricity is the focus of this chapter. It moves from fundamentals to concrete configurations with applications and examples. It focuses on how each storage technique contributes to its own unique time scale and how combining longevity, cost, complexity, and grid services in a unified evaluation structured system [1–4].

3.2 Integration Photovoltaic Power with Energy Storage Systems

There are various factors involved in integrating photovoltaic (PV) power with Energy Storage Systems (ESS). They collectively influence the successful integration of PV power with ESS, improving renewable energy adoption and grid stability. Integration depends on more detailed technical, environmental, and operational aspects as technology, sizing, control strategies, weather variability, safety, … [2, 4, 5].

Figure 3.1 illustrates each component affecting integration separately, in reality, these factors interact. For example, adoption is influenced by cost and policy affects cost.

Technical, environmental, economic, and policy/social are the four primary categories into which elements are classified (Fig. 3.2).

- Performance and efficiency are directly impacted by technical elements including control techniques and system sizing.

D. Rekioua, *Energy Storage for Renewable Energy Systems*, Green Energy and Technology, https://doi.org/10.1007/978-3-032-19589-0_3

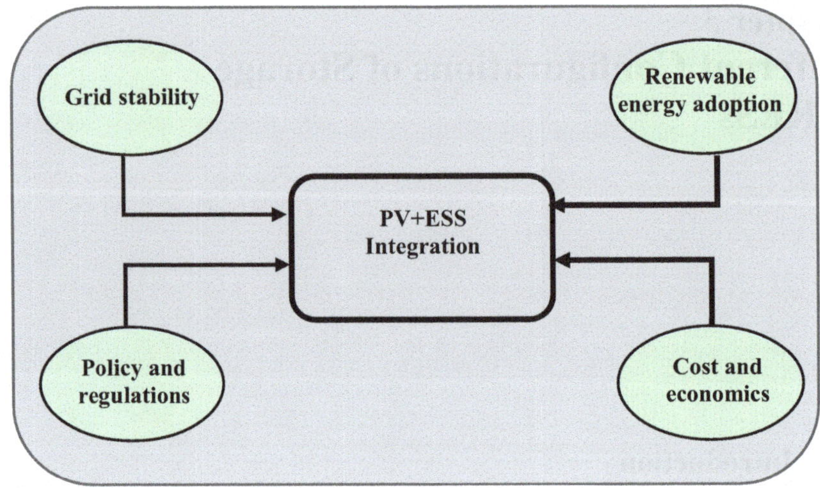

Fig. 3.1 Factors influencing PV + ESS integration

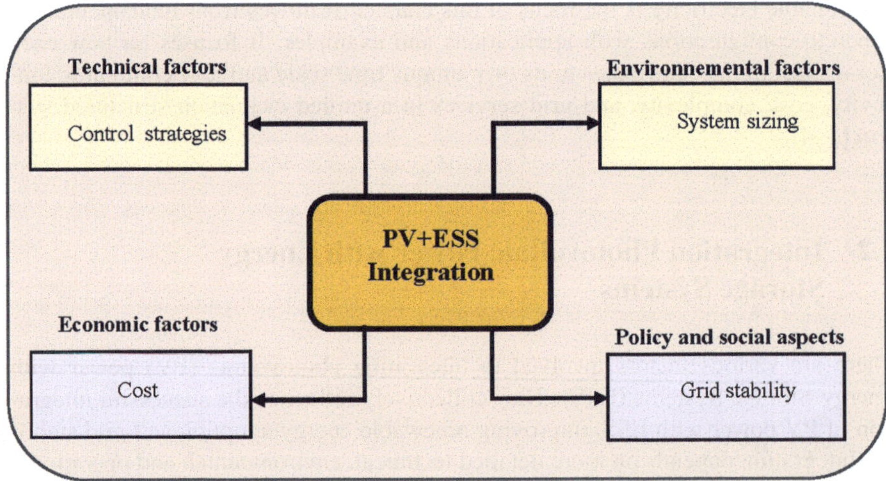

Fig. 3.2 Main factors affecting photovoltaic and energy storage system integration

- Climate and other environmental factors affect the best design of PV and storage components.
- Policy and social aspects, represented by grid stability and regulations, define the legal and operational conditions within which PV + ESS projects operate.
- Economic factors, in particular cost, include investment cost, which have a significant impact on feasibility of the project.

3.3 Integration of Storage in Photovoltaic Power Systems

Photovoltaic power systems can incorporate energy storage in several configurations. Daily energy shifting and backup are provided by single-storage systems, which are simple and cost-effective but only use one technology's advantages. By combining two complementing devices, dual-storage allows the high-energy unit to manage fast transients while the high-power unit manages longer transients. Hybrid (often used to mean two or more) and multi-storage (three or more) architectures extend this idea by combining technologies to cover milliseconds-to-seasonal needs: power-quality and ramp-rate control, daily self-consumption, and long-duration resilience [4–7] (Fig. 3.3).

3.3.1 Single Storage

3.3.1.1 Common Configurations

Typical PV system configurations combined with various ESS types are shown in Fig. 3.4.

1. PV with Lead-Acid batteries: a conventional, affordable, and dependable solution that works well for short-term energy storage (STES).
2. PV with Lithium-Ion batteries: perfect for applications needing frequent cycling and compact storage.
3. PV with supercapacitors: often used to reduce PV variations. This technology offers a high power density and quick charging and discharging for short-term energy storage.
4. PV with hydrogen fuel cells: this system uses electrolysis to turn excess PV electricity into hydrogen for long-term storage, which may then be converted back into electricity for seasonal storage.

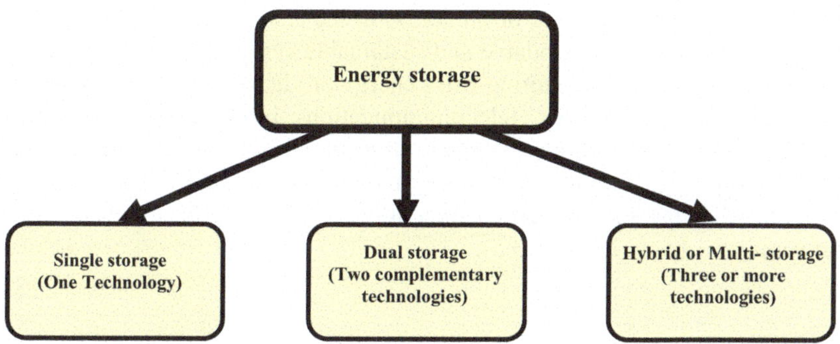

Fig. 3.3 PV storage configurations: single, dual, and hybrid/multi-storage

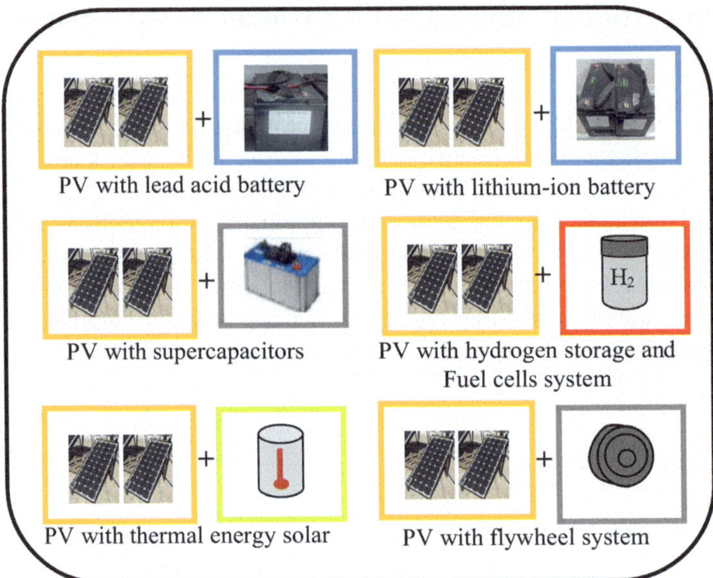

Fig. 3.4 Some configurations of PV systems integrated with storage

5. PV with thermal solar storage: this option combines PV with solar thermal collectors to store heat, allowing for commercial or residential use of a combined heat and power supply.
6. PV with flywheel system: the flywheel is accelerated by converting extra electrical energy from the PV array during times of high irradiance into mechanical energy. The flywheel decelerates to release stored energy to the grid or local load at times of low irradiance or when the load increases.

3.3.1.2 Applications

Isolated locations with irregular or no access to the central electrical grid, PV systems with battery offer a dependable and sustainable system. These systems ensure a steady supply of power by storing extra energy produced. Rural electrification in villages for small appliances, telecommunications infrastructure that supplies remote telecom towers and relay stations, educational institutions that use them to supply electricity to computers, lighting, and communication devices for digital learning, monitoring systems and weather stations in remote locations, emergency power supply, … [4, 6, 7]. (Fig. 3.5).

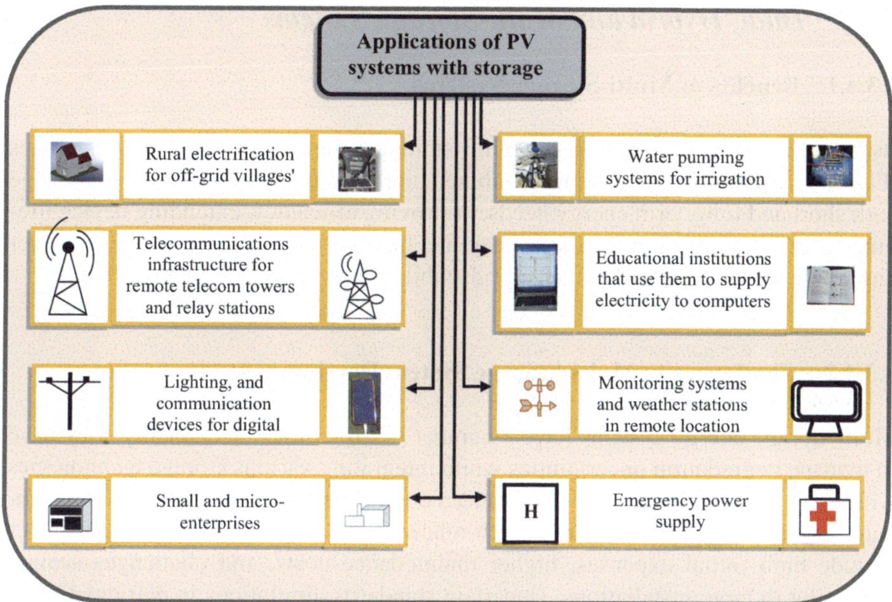

Fig. 3.5 Common applications of PV systems with storage

3.3.2 Hybrid or Multi Energy Storage

A more complete and adaptable solution for RESs is offered by hybrid energy storage systems, which incorporate many technologies. These hybrid systems combine various energy storage technologies to maximize their advantages while limiting their disadvantages [2, 3, 7, 8].

3.3.3 Necessity of Multi-Storage Systems

Renewable energy sources are naturally uncertain because their output depends on environmental factors. ESSs must store extra energy to reduce these variations and obtain a steady energy supply. By combining various storage technologies, multi-storage systems aim to maximize each storage type's advantages while reducing its drawbacks.

3.3.4 *Dual, Hybrid and Multi-Storage Systems*

3.3.4.1 Benefits of Multi-Storage Systems

By combining these technologies, multi-storage systems offer several advantages (Fig. 3.6). Multi-storage systems combine complementary technologies to manage both short and long-term energy needs, improving efficiency, extending device lifetime, and reducing overall costs. They are flexible and scalable, support higher renewable integration, and enhance grid stability, power quality, and resilience.

3.3.4.2 Challenges in Multi-Storage System Development

Multi-storage energy systems have a number of difficulties. Technically, they have to manage degradation uncertainties while integrating various storage technologies with various dynamics, requiring advanced control strategies and maintaining compatibility and efficient interaction with renewables and grids. Economic obstacles include high initial expenses, higher maintenance costs, and challenges scaling from pilot to large installations. Uncertain standards, limitations in participating in multiple grid services, and investment risks resulting from policy changes are examples of regulation and economic challenges [2, 3, 7].

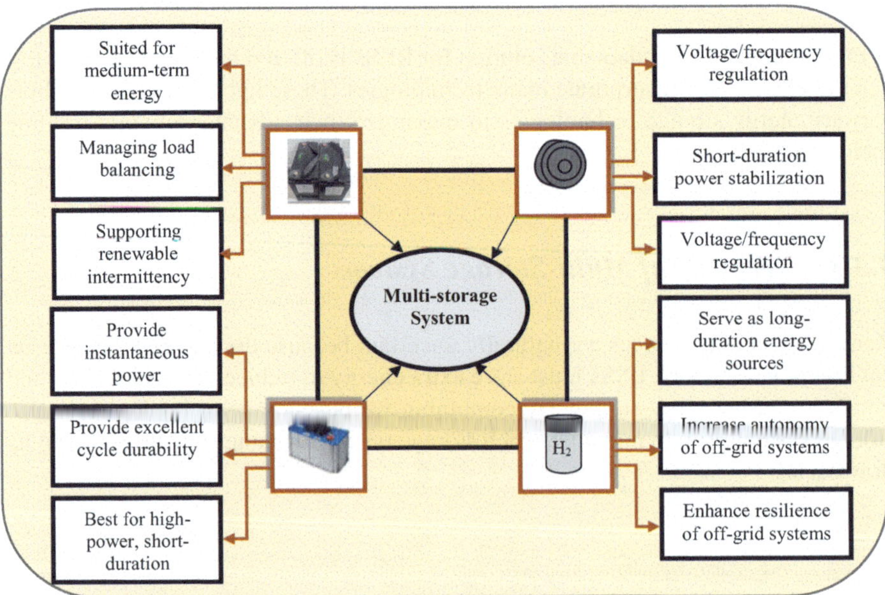

Fig. 3.6 Main advantages and benefits of multi-storage systems

3.3.4.3 Different Configurations

Various configurations are used based on the specifications and the application size. Dual ESSs, such as batteries coupled with supercapacitors, hydrogen, or thermal units, are effective in balancing short-term fluctuations with medium-duration needs, extending battery life and improving overall reliability. Hybrid storage systems integrate three complementary technologies (batteries, supercapacitors, and thermal or flywheels) to optimize performance across multiple timescales, providing fast response, sustained supply, and better grid stability. At a larger scale, multi-storage systems combine several technologies (batteries, supercapacitors, thermal, hydrogen, pumped hydro, etc.) into a coordinated framework, offering maximum flexibility, scalability, and cost-effectiveness while enabling higher penetration of solar energy. Together, these approaches allow PV systems to evolve from simple backup solutions to smart, adaptive energy units capable of supporting both local and large-scale energy demands.

In PV applications, dual storage systems are relatively simple and cost-effective, but their flexibility remains limited since they can only balance two timescales. Hybrid storage systems, by integrating three complementary technologies, achieve higher efficiency and reliability but introduce greater complexity in control and management. Multi-storage systems offer the most powerful and scalable solution (Fig. 3.7).

3.3.4.4 Mathematical Combinations in a Multi-Storage Energy System

Multi-storage systems typically combine several types of energy storage technologies. The number of possible combinations in a multi-storage energy system relies mainly on the number of distinct energy storage technologies you consider and the size of the combinations you want to explore. The number of combined technologies can vary, ranging from 2 to several (3, 4, or even more), depending on the specific needs of the system, on complexity, on compatibility of the technologies, and on the performance objectives. Most systems, however, use 2 to 4 technologies since, over this level, complexity, costs, and management become more challenging. Furthermore, in advanced projects, it is also possible to combine up to 5 or 6 technologies, but this remains quite rare and must be justified by specific requirements. In general, mathematical formulations are employed to systematically determine the different possible combinations of multiple storage technologies, taking into account system constraints and performance objectives. If we have n storage technologies, then the number of possible combinations depends on how many technologies are combined at once (denoted by k_c) [2, 7].

The binomial coefficient indicates how many possibilities there are to select k technology from n:

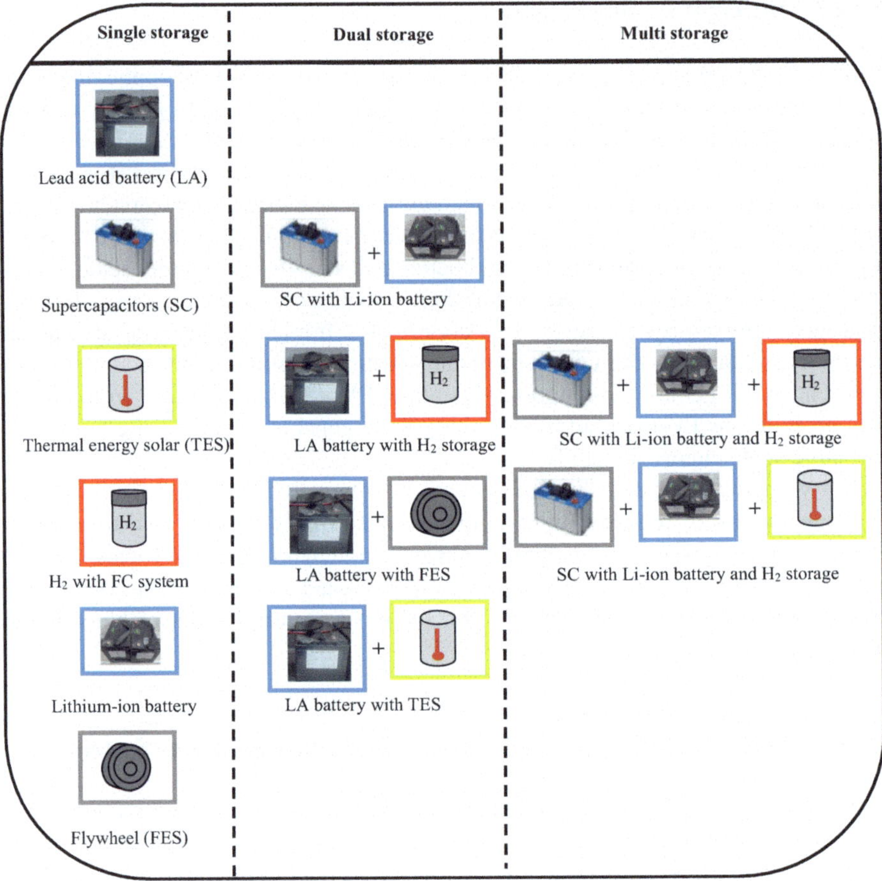

Fig. 3.7 Hierarchical configurations of storage technologies in PV systems

$$C(n_c,k_c)=\binom{n_c}{k_c}=\frac{n_c!}{k_c!(n_c-k_c)} \tag{3.1}$$

We have a set of storage technologies S with size $n = |S|$ and we want all combinations of size $k_c \in \{2,3,4\}$. The conditions will be no singles allowed, so $k_c \neq 1$), no repetition of the same technology in a combination (combinations without replacement) and the order combinations of does not matter. The number of possible combinations of size k from n technologies is given by the binomial coefficient.

The total number of valid hybrid combinations is:

$$C=\sum_{2}^{n_c}\binom{n_c}{k_c}=\binom{n_c}{2}+\binom{n_c}{3}+\ldots.+\binom{n_c}{k_c} \tag{3.2}$$

But we know that the sum of all binomial coefficients is:

$$\sum_{k_c}^{n_c} \binom{n_c}{k_c} = 2^{n_c} \tag{3.3}$$

$$C = 2^{n_c} - n_c - 1 \tag{3.4}$$

where: the total number of subgroups is 2^n (including empty and single ones), subtracting n means removing the single-technology cases, subtracting 1 means removing the empty set.

Table 3.1 summarizes the number of possibilities to select precisely k_c technologies from n:

3.3.4.4.1 Example 1 (See Chap. 8, Sect. 8.2.4.1)

3.3.4.4.2 Commonly Energy Storage Technologies Combinations

The many combinations of the most ESSs used are assembled in Table 3.2. This comprises 55 examples that are doubled (cells in blue color), 11 cases that require no combinations (cells in red color), and novel or upcoming solutions (highlighted in yellow color).

Also, to increase the number of combinations, it will be interesting to add an optimization criterion to select the best combinations based on variables (cost, efficiency...) by adding an objective function to apply to each combination.

An optimization criterion's mathematical expression is predicated on:

Table 3.1 Choice of k_c technology

Cases	k_c	Combinations
Case 1. Only single technologies	$k_c = 1$	$C(n_c,1) = \binom{n_c}{1} = \frac{n_c!}{1!(n_c-1)}$
Case 2. Hybrid of exactly two technologies	$k_c = 2$	$C(n_c,2) = \binom{n_c}{2} = \frac{n_c!}{2!(n_c-2)}$
Case 3. Hybrid of exactly three technologies	$k_c = 3$	$C(n_c,3) = \binom{n_c}{3} = \frac{n_c!}{3!(n_c-3)}$
. .	. .	. .
Case k. hybrid of exactly k technologies	k_c	$C(n_c,k_c) = \binom{n_c}{k_c} = \frac{n_c!}{k_c!(n_c-k_c)}$

Table 3.2 An overview of the various energy storage arrangements

	BES	SES	FES	TES	SMES	PHES	HES	CAES	GES	CES	SCES
BES		BES+ SES	BES+ FES	BES+ TES	BES+ SMES	BES+ PHES	BES+ HES	BES+ CAES	BES+ GES	BES+ CES	BES+ SCES
SES			SES+ FES	SES+ TES	SES+ SMES	SES+ PHES	SES+ HES	SES+ CAES	SES+ GES	SES+ CES	SES+ SCES
FES				FES+ TES	FES+ SMES	FES+ PHES	FES+ HES	FES+ CAES	FES+ GES	FES+ CES	FES+ SCES
TES					TES+ SMES	TES+ PHES	TES+ HES	TES+ CAES	TES+ GES	TES+ CES	TES+ SCES
SMES						SMES+ PHES	SMES+ HES	SMES+ CAES	SMES+ GES	SMES+ CES	SMES+ SCES
PHES							PHES+ HES	PHES+ CAES	PHES+ GES	PHES+ CES	PHES+ SCES
HES								HES+ CAES	HES+G ES	HES+ CES	HES+ SCES
CAES									CAES+ GES	CAES+ CES	CAES+ SCES
GES										GES+ CES	GES+ SCES
CES											CES+ SCES
SCES											

Let: $S = \{s_1, s_2, \ldots, s_n\}$ be the set of available technologies, and C_k be the set of combinations without repetition of size k ($k = 2$ or $k = 3$). Each technology s_i is characterized by a vector of criteria (for example cost c_i, efficiency e_i, lifetime, carbon footprint, etc.) or scalar variables.

For a combination $c = \{s_{i1}, s_{i2}, \ldots, s_{ik}\} \in C_{kc}$, we define an objective function $f(c)$ representing the overall performance of the combination c.

A simple example, weighting several criteria, could be:

$$f(c) = w_c . \sum_{j=1}^{k} c_{ij} + w_e . \sum_{j=1}^{k} e_{ij} + \ldots \tag{3.5}$$

Where: $w_c, w_e, \ldots$ are weights defining the relative importance of each criterion.

Depending on the case, the goal may be to minimize (total cost) or maximize (total efficiency). Given the possible combinations $C = U_{k=2}^{3} . C_k$, we obtain c^* which is the combination (from the set C of all possible combinations) that optimizes the objective function $f(c)$. The function f could represent performance, cost, efficiency, or a weighted combination so, the objective can be maximization or minimization.

$$c^* = argmaxf(c) \text{ or } c^* = arg\min f(c) \tag{3.6}$$

3.3.4.4.3 Example 2 (See Chap. 8, Sect. 8.2.4.2)

3.3.4.5 Applications for Dual, Hybrid and Multi Storage Technologies Used in PV Systems

3.3.4.5.1 Photovoltaic/Fuel Cell/Battery (PV/FC/BT)

In this application, an electric vehicle (EV) is powered by the PV/FC/BT system. The proposed power system with dual storage ensures continuous and reliable energy production by integrating RESs with dual storage technologies: battery bank and hydrogen storage through a proton exchange membrane fuel cell (PEMFC) [7, 9–11]. The PV generator and the PEMFC operate in parallel (via DC/DC converters), while the battery provides fast-response storage for short-term fluctuations, and the hydrogen–fuel cell unit ensures sustained energy delivery during extended demand or low solar generation [12–15]. The modelling of each subsystem is given in the Table 3.3.

The resistive force is:

$$F_r = F_{tire} + F_{aero} + F_{slope} \tag{3.7}$$

With the various forces are shown in Table 3.4:

3.3.4.5.2 Photovoltaic/Battery/Supercapacitor (PV/BT/SC)

When batteries and SCs are combined, their complementing benefits are maximized, creating a more dependable and affordable storage option. Efficient energy management supervision (EMS) is necessary (Fig. 3.8).

Thus [16–20]:

$$\begin{cases} P_{loadcalc} = P_{PV} + \left(P_{SC} + P_{batt}\right) \\ \Delta P = P_{load} - P_{PV} \end{cases} \tag{3.8}$$

Table 3.5 provides a summary of the several equations that were used to create the algorithm..

Based on these equations, and by applying an effective algorithm eleven cases can be available (Fig. 3.9).

First, the storage (supercapacitor or battery) is already full in cases 1 and 2, thus it is isolated to avoid overcharging. In case 3, both storage units are disconnected to prevent ineffective charging if PV alone can support the load when they are both empty. In cases 4 and 5, PV generation surpasses the load. As a result, the excess energy initially charges the battery (case 4) or the SC (case 5), while PV continues

Table 3.3 Main equations used in the model PV/FC/BT

$P_{PV} + P_{FC} + P_{batt} = P_{EV}$	
PV system	$P_{PV} = V_{PV} \cdot I_{PV}$ $I_{PV} = I_{ph} - I_d - I_{R_{sh}}$ $I_{PV} = I_{ph} - I_0 \left[\exp\left(\frac{q \times V_{PV} + R_s \times I_{PV}}{A \times N_s \times K \times T_j} \right) - 1 \right] - \frac{V_{PV} + R_s \times I_{PV}}{R_{sh}}$
Hydrogen storage –PEMFC	$P_{FC} = V_{FC} \times I_{FC}$ $V_{FC} = E_{Nerst} + U_{act} - U_{Ohm} - U_{conc}$ $V_{FC} = \alpha_1 + \alpha_2 \left(T_{FC} - 298.15 \right) + \alpha_3 . T_{FC} \left(0.5 \ln P_{O_2} + \ln P_{H_2} \right)$ $+ \beta_1 + \beta_2 . T_{FC} + \beta_3 . T_{FC} . \ln(j) + \beta_4 . T_{FC} . \ln\left(C_{O_2} \right)$
Batteries	$V_{batt} = n_b . E_b \pm n_b . R_{batt} . I_{batt}$ $V_{batt-charge} = n_b . \left[2 + 0.16 SOC \right]$ $+ n_b . \frac{I_{batt}}{C_{10}} \left[\frac{6}{1 + I_{batt}{}^{1.3}} + \frac{0.27}{SOC^{1.5}} + 0.002 \right] \left(1 - 0.007." \, T \right)$ $V_{batt-discharge} = n_b . \left[1.965 + 0.12 SOC \right]$ $- n_b . \frac{I_{batt}}{C_{10}} \left[\frac{4}{1 + I_{batt}{}^{1.3}} + \frac{0.27}{SOC^{1.5}} + 0.002 \right] \left(1 - 0.007." \, T \right)$ $C_{batt} = C_{10} . \frac{1.67 \left(1 + 0.005" \, T \right)}{1 + 0.67 \left(\frac{I}{I_{10}} \right)^{0.9}}$ $SOC = 1 - \frac{I_{batt} . t}{C_{batt}}$
DC/DC converters:	$V_{out} = \frac{V_{in}}{1 - D}$ $P_{batt} = V_{batt} . I_{batt} = \eta_{conv} . V_{batt} . I_{batt}$
Electric vehicle	$P_{PV} + P_{FC} + P_{batt} = P_{EV} + P_{losses}$
Control strategies	Coordination between PV, battery, and fuel cell to optimize energy usage Control algorithms

Table 3.4 Different forces on vehicule with its equations

Forces		Equations
F_{tire}	Rolling resistance force	$F_{tire} = m.\ g.\ f_{ro}$
F_{aero}	Aerodynamic drag force	$F_{aero} = \frac{1}{2}\varphi_{air}.A_f.C_d.V^2$
F_{slope}	Climbing force	$F_{slope} = m.\ g.\ \sin(\beta)$

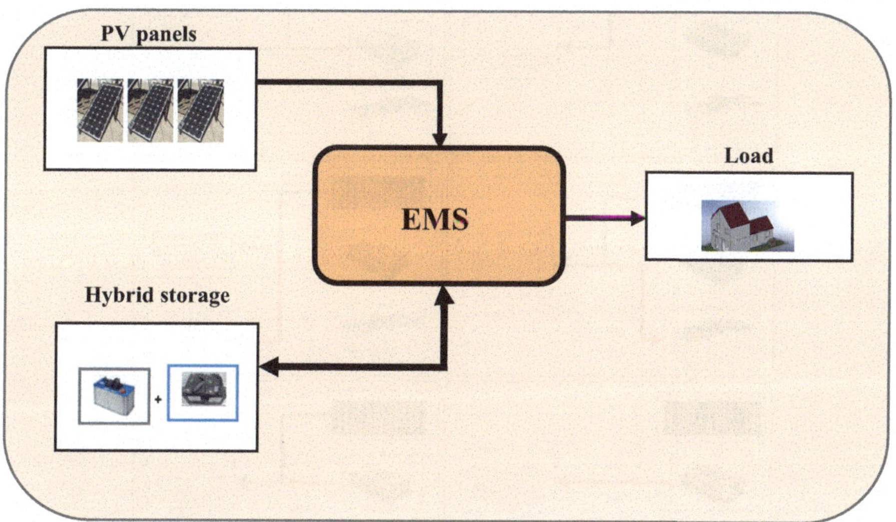

Fig. 3.8 Autonomous PV system with dual storage

Table 3.5 The different equations used in the algorithm

Modes	Equations
Case 1	$\Delta P = 0 \quad P_{load} = P_{PV} \quad SOC_{batt} \geq SOC_{batt-max}$
Case 2	$\Delta P = 0 \quad P_{load} = P_{PV} \quad SOC_{SC} \geq SOC_{SC-max}$
Case 3	$\Delta P = 0 \quad P_{load} = P_{PV} \quad SOC_{batt} < SOC_{batt-min} \quad SOC_{SC} < SOC_{SC-min}$
Case 4	$\Delta P > 0 \quad P_{load} = P_{PV} \quad SOC_{batt} \leq SOC_{batt-min}$
Case 5	$\Delta P > 0 \quad P_{load} = P_{PV} \quad SOC_{SC} \leq SOC_{SC-min}$
Case 6	$\Delta P > 0 \quad P_{load} = P_{PV} \quad SOC_{batt} \leq SOC_{batt-min} \quad SOC_{SC} \leq SOC_{SC-min}$
Case 7	$\Delta P < 0 \quad P_{load} = P_{batt} \quad P_{PV} = 0 \quad SOC_{batt} > SOC_{batt-min}$
Case 8	$\Delta P < 0 \quad P_{load} = P_{PV} + P_{batt} \quad P_{PV} \neq 0 \quad SOC_{batt} > SOC_{batt-min}$
Case 9	$\Delta P < 0 \quad P_{load} = P_{SC} \quad P_{PV} = 0 \quad SOC_{bSC} > SOC_{SC-min}$
Case 10	$\Delta P < 0 \quad P_{load} = P_{PV} + P_{SC} \quad P_{PV} \neq 0 \quad SOC_{SC} > SOC_{SC-min}$
Case 11	$\Delta P < 0 \quad P_{load} = 0 \quad P_{PV} \neq 0 \quad SOC_{batt} \leq SOC_{batt-min} \quad SOC_{SC} \leq SOC_{SC-min}$

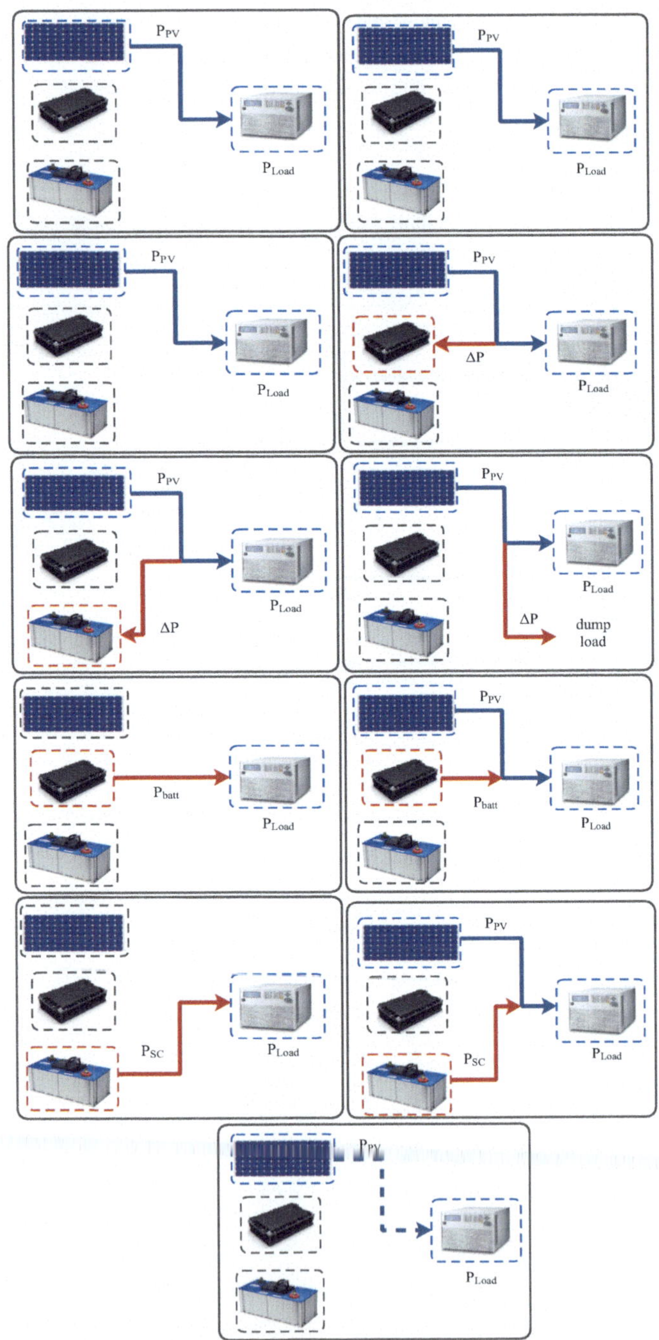

Fig. 3.9 Different cases in PV/Battery/SCs

to meet the load. Furthermore, in case 6, the remaining excess is sent to a dump load if PV still surpasses the load even after both storage units are full. When PV is insufficient in cases 7 and 8, the battery then covers up the deficit, either by itself in case 7 when PV is zero or in combination with PV in case 8 when PV is nonzero. In the same way, cases 9 and 10 occur when the SC is charged but the battery is not available. In these cases, the SC provides the load, either alone (if P_{PV} is zero) or with PV (if P_{PV} is nonzero). When PV cannot meet the load and both storage units are full, the storage is disconnected and the available PV cannot generate electricity to meet the demand. As a result, load shifts or external backup is needed in case 11.

3.4 Integration of Storage in Wind Turbine Systems

Wind energy is intermittent and fluctuating, which makes the integration of storage systems essential for ensuring stable and reliable power supply. Different storage technologies are employed to complement wind turbines: batteries provide medium-term balancing and energy management, SCs manage fast transients and protect batteries from stress, while long-term solutions as hydrogen, CAES, or pumped hydro enable seasonal or large-scale storage. Increasingly, interdisciplinary approaches are adopted to address various aspects of energy storage, combining insights from electrical engineering, materials science, and control optimization. In this context, dual, hybrid, and multi-storage configurations are widely explored in wind turbine systems to smooth power output, enhance grid stability, reduce storage units degradation, and achieve higher overall system efficiency and reliability [2, 3, 7, 10].

3.4.1 Interdisciplinary Approaches to Address Various Aspects of ESSs

Storage studies in wind power systems often involve interdisciplinary approaches to address various aspects of energy storage integration and optimization. These approaches combine expertise from electrical engineering (power electronics, control strategies, grid integration), mechanical engineering (aerodynamics, drivetrain dynamics, structural loading), materials science (advanced electrodes, hydrogen storage materials, thermal fluids), and computer science/mathematics (optimization, machine learning, predictive modeling). In addition, economics and policy/regulation play a central role through cost, benefit analysis, market mechanisms, business models for storage deployment. Such interdisciplinary work is crucial to solve challenges like intermittency, system stability, lifetime degradation of storage devices, and cost-effectiveness (Fig. 3.10.)

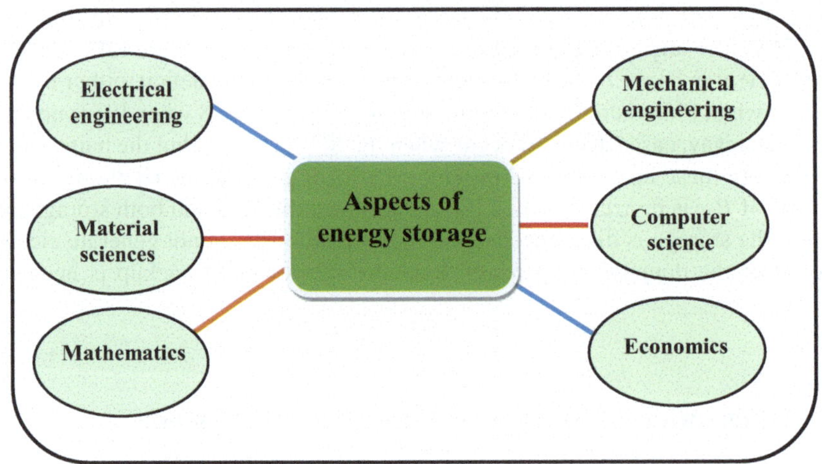

Fig. 3.10 Interdisciplinary approaches to address various aspects of energy storage

3.4.2 *Integration of WTb with ESSs*

The integration of WTb systems with ESSs involves several factors that properly taken into account to obtain efficiency, reliability, and sustainability. By integrating ESS, the control strategy balances this variability to obtain a stable and controllable power to either an autonomous load (off-grid) or the grid.

- During charging mode, ESS absorbs the surplus wind power: If $P_{WTb} \rangle P_{load}$ then $P_{ESS}^{char} = P_{WTb} - P_{load}$
- During discharging mode: If $P_{WTb} \langle P_{load}$ then $P_{ESS}^{disch} = P_{load} - P_{WTb}$
- And during idle mode where ESS remains inactive: $P_{load} = P_{WTb}$

3.4.3 *Single Storage in WTbs*

A number of ESSs are frequently applied in wind turbines (WTbs) to improve flexibility and reliability (Fig. 3.11). Examples of these include wind turbines that have batteries, supercapacitors, flywheels, hydro storage, CAES, hydrogen storage, and more [21–27].

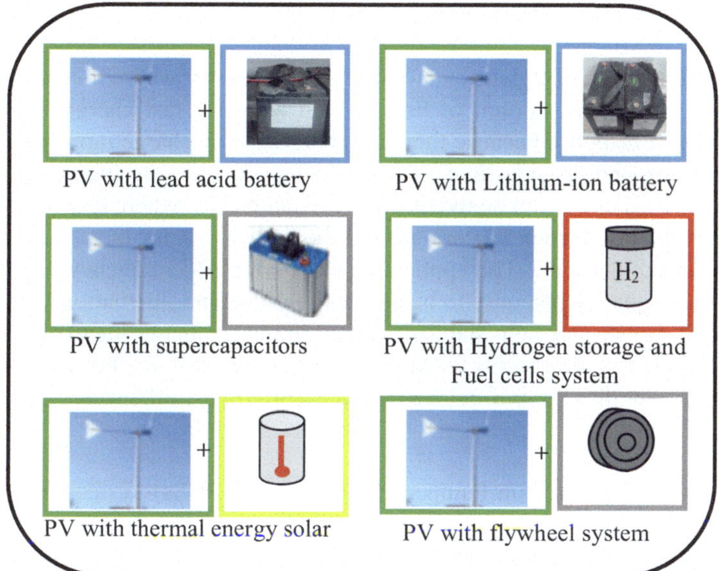

Fig. 3.11 Examples of single energy storage systems integrated with WTbs

3.4.3.1 Common Configurations of Wind Turbine with Single Storage

3.4.3.1.1 Wind Turbine System with Batteries

The storage can be Lithium-ion, lead-acid, or sodium-sulfur (NaS). It's a mature technology, scalable from small microgrids to utility-scale wind farms, relatively high efficiency, but some challenges as limited lifetime due to cycling, thermal management need, environmental concerns, ….. The application of this system are on microgrids to provide reliable power for off-grid or remote communities, autonomous loads to maintain power quality and availability for critical infrastructure, frequency regulation, backup power because batteries supply energy during outages or grid disturbances, …….

3.4.3.1.2 Wind Turbine System with Supercapacitors (SCs)

In this case SCs offers distinct advantages primarily for managing very short-term fluctuations and enhancing power quality. They are extremely fast charge/discharge making them ideal for smoothing transient fluctuations caused by wind turbulence, very high cycle life without significant degradation. The most challenge can be in low energy density (not suited for long-term storage). The main applications are for power quality, stabilizing turbine pitch control systems, and supporting frequency regulation.

3.4.3.1.3 Wind Turbine System with Flywheels

The most advantages of flywheels are high power density where they can provide large power in a short amount of time, which ideal for fast energy injection or absorption, fast response because they can charge and discharge rapidly to smooth wind variability and provide immediate grid support, and long lifetime with little degradation. Challenges can be the mechanical complexity where flywheels require precise engineering in rotors, bearings, and the higher cost for large-scale systems due to advanced materials. The application can be for grid frequency regulation to provide fast frequency response to balance supply-demand fluctuations, smoothing wind variability because help stabilize power output, reducing fluctuations caused by wind turbulence, short-term backup in autonomous systems because flywheels can supply power during brief outages or transient grid disturbances.

3.4.3.1.4 WTb System with Pumped Hydro Storage (PHSs) Systems

PHSs pump water to an elevated reservoir during excess wind generation, release through turbines when demand is higher. The main advantages are very large-scale and long-duration storage, making it ideal for balancing intermittent wind power on a grid scale, mature technology and cost-effective at utility scale. Needs reservoirs, elevation difference and environmental impacts can be some challenges without forget the high initial investment. Applications include grid-scale wind integration, seasonal balancing, …….

3.4.3.1.5 Wind Turbine System with CAES

Large storage capacity, can provide several hours to days of backup, relatively low cost compared to some other large-scale storage technologies, are the most advantages of CAES. However, it requires suitable geology for storage caverns, lower round-trip efficiency because energy is lost due to heat generated during air compression and cooling during decompression, needs hybridization with thermal storage to improve efficiency. The applications are for grid-level integration of large wind farms, medium/long-term balancing.

3.4.3.1.6 Wind Turbine System with Hydrogen Storage

Electrolysis uses electricity provided by wind to create H_2, which is stored and either used as fuel or converted back into energy using fuel cells. Because hydrogen may be produced and stored for extended periods of time, it is a seasonal storage method, enabling energy availability when wind generation is low or during seasonal variations, a versatile use (as electricity, transport, and industry), and a potential for decarbonization because hydrogen derived from wind energy is green and can replace fossil fuels in different sectors.

Some challenges can be around low efficiency because there are energy losses during electrolysis, compression/storage, and conversion back to power, expensive infrastructure because electrolyzers, hydrogen storage tanks (compressed or liquid), fuel cells, and distribution networks require high capital investments, hydrogen handling storage issues because it requires careful management in storage, transport, and usage to prevent accidents. Applications are for large-scale renewable integration, sector coupling (power-to-gas, power-to-industry), long-duration storage.

3.4.3.2 Applications

3.4.3.2.1 Wind Turbine with Battery Storage (See Chap. 8, Sect. 8.2.5)

This application supports wind energy integration at small scale for residential or remote grid-connected use cases with battery energy storage (Fig. 3.12).

3.4.3.2.2 Wind Turbine with Flywheel Storage (See Chap. 8, Sect. 8.2.6)

This application supports wind energy integration at small scale for residential or remote grid-connected use cases with flywheel energy storage (Fig. 3.13).

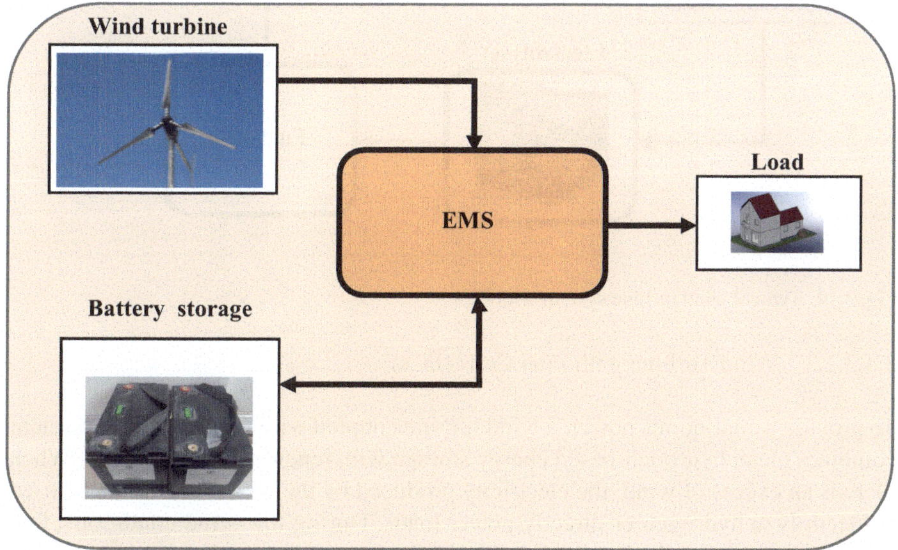

Fig. 3.12 Typical wind turbine system with battery storage

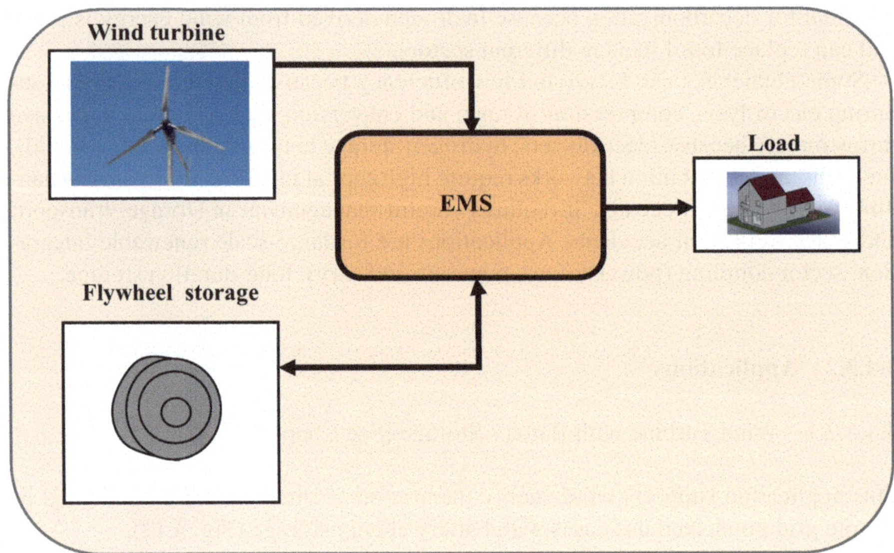

Fig. 3.13 Typical wind turbine system with flywheel storage

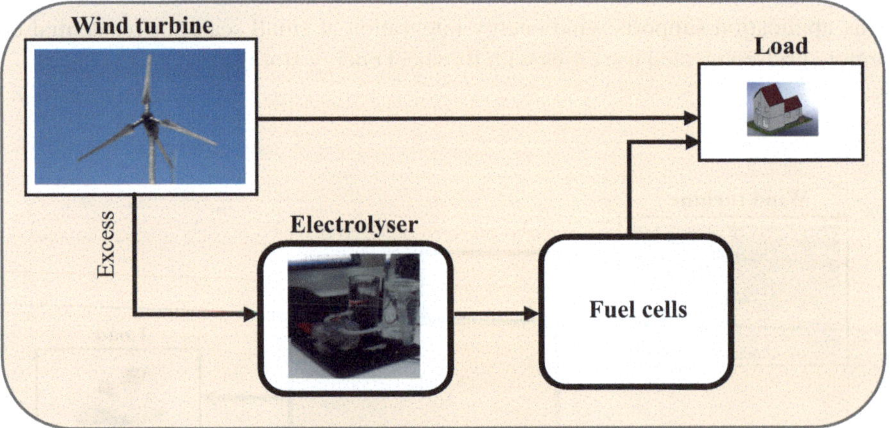

Fig. 3.14 Typical wind turbine system with FCs

3.4.3.2.3 Wind Turbine with Fuel Cells (FCs)

To provide a continuous power, a wind turbine coupled with fuel cells (FCs) system combines clean hydrogen-based energy storage with renewable wind energy. When there is an excess of wind, the electricity produced by the wind turbine can be used to electrolyze hydrogen or directly power loads. During low wind conditions, fuel cells then transforms the stored hydrogen back into power, increasing system autonomy and dependability (Fig. 3.14).

3.4.4 *Dual, Hybrid and Multi-Storage in WTbs*

The choice of a multi-storage system is influenced by various factors [3, 7, 10, 11].

3.4.4.1 The Most Various Cases of Multi-Storage in WTbs

Table 3.6 summarizes the most common multi-storage systems employed in WTbs. They provide a number of advantages, however, each combination's benefit may differ based on the particular application and wind system needs.

They also have a number of drawbacks (Table 3.7).

Cost refers to the system's higher total cost, which requires multiple components and complex control systems. Combining several energy storage methods results in complexity, and needs the use of advanced control and monitoring systems to ensure correct operation. Maintenance is a challenge since the many parts of a storage system may require additional attention and maintenance, which could increase costs and cause difficulties (Fig. 3.15).

3.4.4.1.1 Wind Turbine/Batteries/Supercapacities

This application suggests a hybrid storage solution that combines SCs and batteries. An efficient energy management algorithm is required to control the two storage methods (Fig. 3.16).

Table 3.6 Multi storage systems used in wind turbine systems

Multi-storage	STES	LTES	EC	SE	EI	R	F
Flywheel and batteries	H	M	H	L	H	H	H
TES and FES	H	H	H	M	H	M	H
CAES and FES	H	H	H	M	H	L	H
Li-ion batteries and H_2 fuel cells	H	H	H	L	H	L	H
TES and H_2 fuel cells	H	H	H	L	H	L	H
Ultracapacitors and flow batteries	M	H	H	L	H	H	M
Pumped hydro with batteries	H	H	H	L	H	H	H
FES and H_2 fuel cells	H	H	H	L	H	M	H
CAES with TES	H	H	H	M	H	L	H
Battery and CAES	H	H	H	M	H	H	H
Battery, TES, and CAES	H	H	H	L	H	H	H

Table 3.7 Multi-storage WTb's drawbacks

Multi-Storage	Cost	Complexity	Maintenance	Space Requirements
FES and batteries	H	H	M	H
TES and FES	H	H	L	H
CAES and FES	H	H	M	H
Lithium-ion batteries and HES	H	H	M	M
TES and HES	H	H	M	H
Ultracapacitors and flow batteries	H	M	L	M
PHES with batteries	H	H	H	H
FES and HES	H	H	M	H
CAES with TES	H	H	L	H
Battery and CAES	H	H	L	H
Battery and TES	H	H	L	H
Battery, FES, and CAES	H	H	H	H

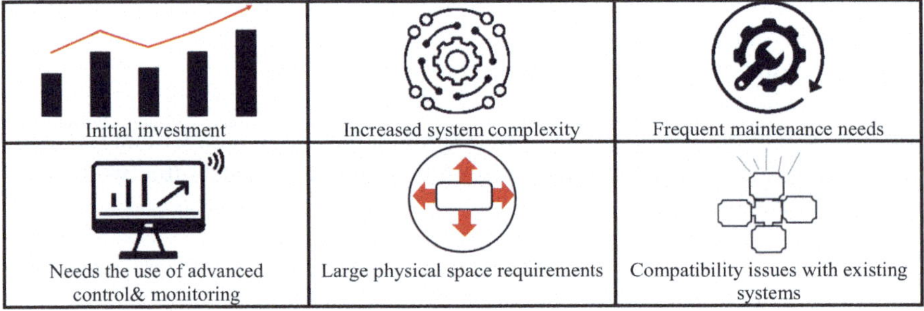

Fig. 3.15 Challenges and limitations of ESS integration

3.4.4.1.2 Wind Turbine/ Batteries/Flywheels

The benefits of mechanical and electrochemical energy storage are combined in a wind turbine system to improve power quality. In this system, the battery controls medium-term energy management and provides a power supply there is not enough wind, while the flywheel responds quickly to short-term power variations, stabilizing voltage and frequency (Fig. 3.17).

3.4.4.1.3 Wind Turbine/ Batteries Fuel Cells

The benefits of both STES and LTES are combined in a WTb system with batteries and GCs to provide a dependable power source even when faced with varying wind conditions. The main source is the wind turbine, which produces power when wind speeds are high enough. The fuel cell technology provides consistent power over prolonged periods of high demand or insufficient wind, while the battery bank controls short-term fluctuation and offers fast-response energy storage. An electrolyzer

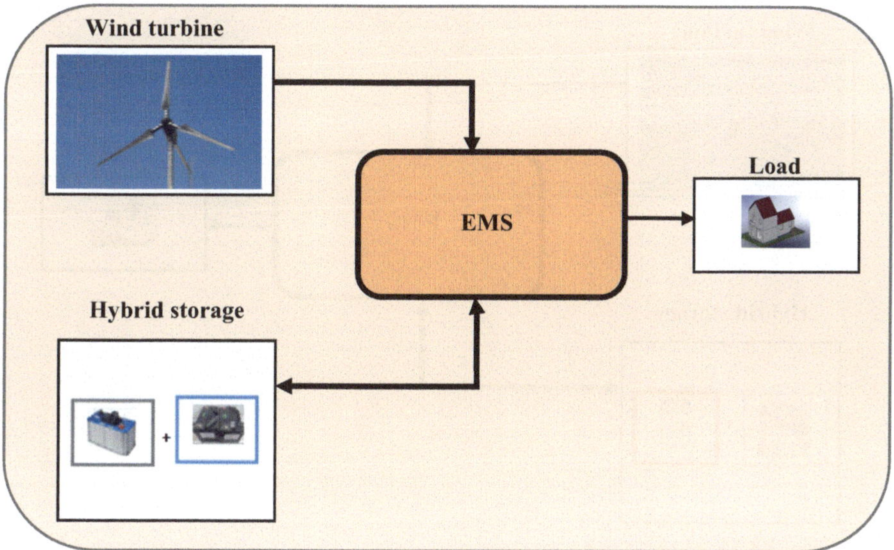

Fig. 3.16 Example of wind turbine with battery and SCs

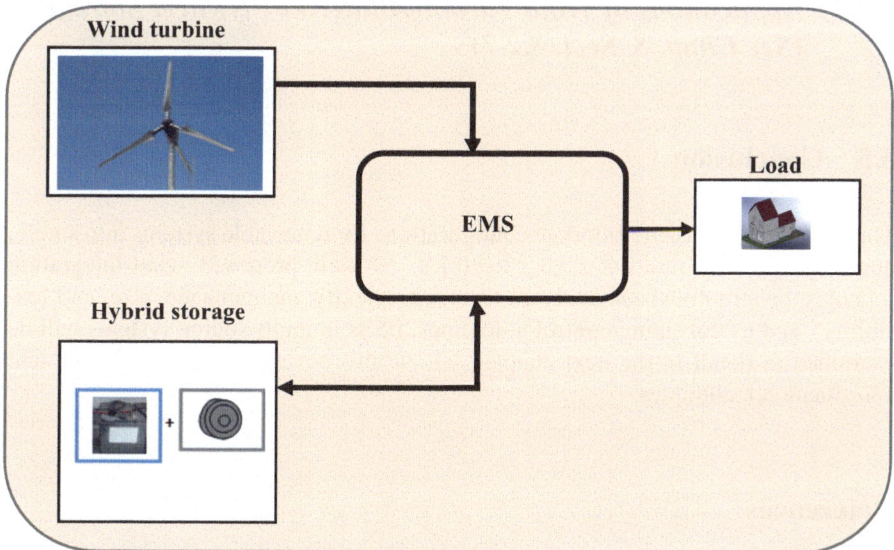

Fig. 3.17 Example of wind turbine with battery and flywheel

uses the wind turbine's excess energy to create hydrogen, which is then transformed back into electricity by the fuel cell when it's needed (Fig. 3.18). For off-grid or remote applications, this hybrid configuration promotes continuous, sustainable energy generation, lowers battery cycle stress, and increases system autonomy.

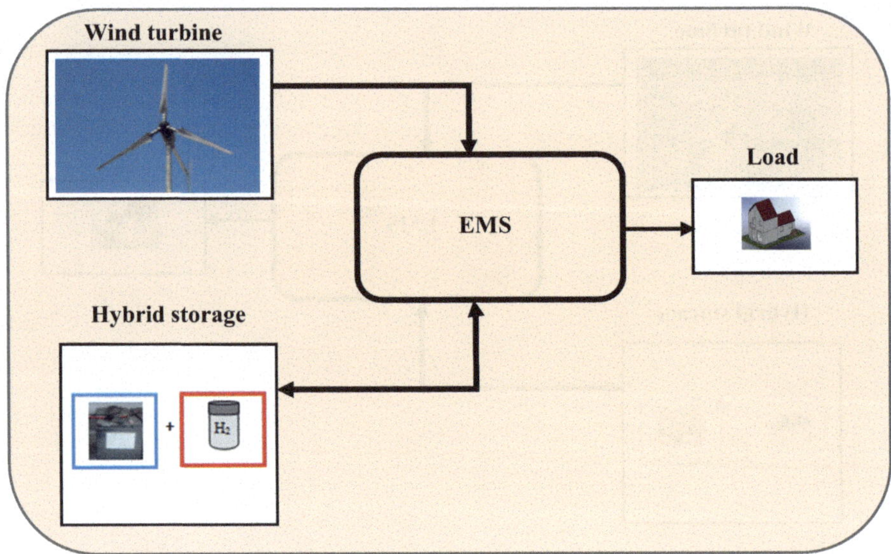

Fig. 3.18 Example of wind turbine with battery and FCs

3.4.5 Applications of Wind Turbine/Batteries/Flywheel Storage (See Chap. 8, Sect. 8.2.7)

3.5 Conclusion

This chapter has classified storage configurations for renewable systems into single, dual, hybrid, and multi-storage categories, as well proposed wind-integration schemes, benefit/drawback analyses (cost, complexity, maintenance, size, and feasibility), and useful sizing/control guidelines. ESSs in multi-source systems will be examined in detail in the next chapter, which also covers control topologies and coordination techniques.

References

1. Abderezzak B, Rekioua D, Binns R, Busawon K, Hinaje M, Douine B et al (2020) Technical feasibility assessment of a PEM fuel cell refrigerator system. Int J Hydrog Energy 45(19):11211–11219. https://doi.org/10.1016/j.ijhydene.2018.04.060
2. Adjati A, Rekioua T, Rekioua D (2021) Use of the dual stator induction machine in photovoltaic–wind hybrid pumping. J Europ des Syst Automat 54(1):115–124. https://doi.org/10.18280/jesa.540113

3. Ahmed S, D'Angola A (2025) Energy storage systems: scope, technologies, characteristics, Progress, challenges, and future suggestions—renewable energy community perspectives. Energies 18(11):2679. https://doi.org/10.3390/en18112679
4. Aissou R, Rekioua T, Rekioua D, Tounzi A (2016) Application of nonlinear predictive control for charging the battery using wind energy with permanent magnet synchronous generator. Int J Hydrog Energy 41(45):20964–20973. https://doi.org/10.1016/j.ijhydene.2016.05.249
5. Aissou R, Rekioua T, Rekioua D, Tounzi A (2016) Robust nonlinear predictive control of permanent magnet synchronous generator turbine using Dspace hardware. Int J Hydrog Energy 41:21047–21056
6. Belaid S, Rekioua D, Oubelaid A, Ziane D, Rekioua T (2022) A power management control and optimization of a wind turbine with battery storage system. J Energy Storage 45:103613. https://doi.org/10.1016/j.est.2021.103613
7. Hassani H, Rekioua D, Zaouche F, Bacha S (2019) Supervision of hybrid renewable energy systems. In: 2019 1st international conference on sustainable renewable energy systems and applications (ICSRESA). IEEE, New York, pp 1–5. https://doi.org/10.1109/ICSRESA49121.2019.9182478
8. Akbari H, Browne MC, Ortega A, Huang MJ, Hewitt NJ, Norton B, McCormack SJ (2019) Efficient energy storage technologies for photovoltaic systems. Sol Energy 192:144–168. https://doi.org/10.1016/j.solener.2018.03.052
9. Horzela-Miś A, Semrau J (2025) The role of renewable energy and storage technologies in sustainable development: simulation in the construction industry. Front Energy Res 13:1. https://doi.org/10.3389/fenrg.2025.1540423
10. Kakouche K, Oubelaid A, Mezani S, Rekioua D, Rekioua T (2023) Different control techniques of permanent magnet synchronous motor with fuzzy logic for electric vehicles: analysis, modelling, and comparison. Energies 16(7):3116. https://doi.org/10.3390/en16073116
11. Laajimi M, Go YI (2019) Energy storage system design for large-scale solar PV in Malaysia: technical and environmental assessments. J Energy Storage 26:100984. https://doi.org/10.1016/j.est.2019.100984
12. Malik FH, Hussain GA, Alsmadi YMS, Haider ZM, Mansoor W, Lehtonen M (2025) Integrating energy storage technologies with renewable energy sources: a pathway toward sustainable power grids. Sustainability 17:4097. https://doi.org/10.3390/su17094097
13. Ould-Amrouche S, Rekioua D, Hamidat A (2010) Modelling photovoltaic water pumping systems and evaluation of their CO_2 emissions mitigation potential. Appl Energy 87(11):3451–3459. https://doi.org/10.1016/j.apenergy.2010.05.021
14. Rahrah K, Rekioua D, Rekioua T, Bacha S (2015) Photovoltaic pumping system in Bejaia climate with battery storage. Int J Hydrog Energy 40(39):13665–13675. https://doi.org/10.1016/j.ijhydene.2015.04.048
15. Rekioua D (2020) MPPT methods in hybrid renewable energy systems. In: Hybrid renewable energy systems. Springer, Cham, pp 79–138. https://doi.org/10.1007/978-3-030-34021-6_3
16. Rekioua D (2020) Power electronics in hybrid renewable energies systems. In: Hybrid renewable energy systems. Springer, Cham, pp 39–77. https://doi.org/10.1007/978-3-030-34021-6_2
17. Rekioua D (2023) Energy storage systems for photovoltaic and wind systems: a review. Energies 16:3893. https://doi.org/10.3390/en16093893
18. Rekioua D (2024) Wind power electric systems: modeling, simulation and control. Springer. https://doi.org/10.1007/978-3-031-52883
19. Rekioua D, Bensmail S, Bettar N (2014) Development of hybrid photovoltaic–fuel cell system for stand-alone application. Int J Hydrog Energy 39:1604–1611. https://doi.org/10.1016/j.ijhydene.2013.03.040
20. Rekioua D, Kakouche K, Babqi A, Mokrani Z, Oubelaid A, Rekioua T, Azil A, Ali E, Alaboudy AHK, Abdelwahab SAM (2023) Optimized power management approach for photovoltaic systems with hybrid battery–supercapacitor storage. Sustainability 15(19):14066. https://doi.org/10.3390/su151914066

21. Rekioua D, Matagne E (2012) Optimization of photovoltaic power systems: modeling, simulation and control. Springer, Cham. https://doi.org/10.1007/978-1-4471-2403-05
22. Rekioua D, Mezzai N, Mokrani Z et al (2024) Effective optimal control of a wind turbine system with hybrid energy storage and hybrid MPPT approach. Sci Rep 14:30013. https://doi.org/10.1038/s41598-024-78847-9
23. Rekioua D, Mokrani Z, Kakouche K et al (2024) Coordinated power management strategy for reliable hybridization of multi-source systems using hybrid MPPT algorithms. Sci Rep 14:10267. https://doi.org/10.1038/s41598-024-60116-4
24. Rekioua D, Mokrani Z, Rekioua T (n.d.) Control of fuel cells–electric vehicle based on direct torque control. Turk J Electromech Energy 3(2):9
25. Rekioua D, Roumila Z, Rekioua T (2008) Étude d'une centrale hybride photovoltaïque–éolien–diesel. J Renew Energ 11(4):623–633. https://doi.org/10.54966/jreen.v11i4.112
26. Xi X, Hu Y, Kang Q, Nie J (2012) A two-level energy storage system for wind. Procedia Environ Sci 12:136. https://doi.org/10.1016/j.proenv.2012.01.257
27. Zeidan M, Al-soud M, Dmour M, Alakayleh Z, Al-qawabah S (2023) Integrating a solar PV system with pumped hydroelectric storage at the Mutah University of Jordan. Energies 16:5769. https://doi.org/10.3390/en16155769

Chapter 4
Storage Systems Used in Hybrid Renewable Energy Systems

4.1 Introduction

Hybrid renewable energy systems (RESs), which usually combine photovoltaic, wind, biomass, and hydro power, are naturally variable. Since each source varies with time, it is unrealistic to assume reliability without storage. Storage balances intermittence, stabilizes voltage and frequency, and reduces generator stress by absorbing excess generation and releasing it when primary resources decline. Higher grid autonomy (or fully islanded operation), better power quality and continuity, and increased energy security against weather, price shocks, or interruptions are the specific benefits [1–5].

4.2 Why Storage Is Needed and What It Does in Hybrid RESs?

4.2.1 Variability in Hybrid RESs

Hybrid renewable energy sources are intermittent and do not naturally follow demand. Through the absorption of excess generation, storage regulates voltage and frequency, minimizes losses, and increases autonomy [6–10].

4.2.2 Why Storage Is Needed in Hybrid RESs?

It balances power supply, smooths fluctuations, ensures efficient switching between power sources [6–10] (Fig. 4.1).

D. Rekioua, *Energy Storage for Renewable Energy Systems*, Green Energy and Technology, https://doi.org/10.1007/978-3-032-19589-0_4

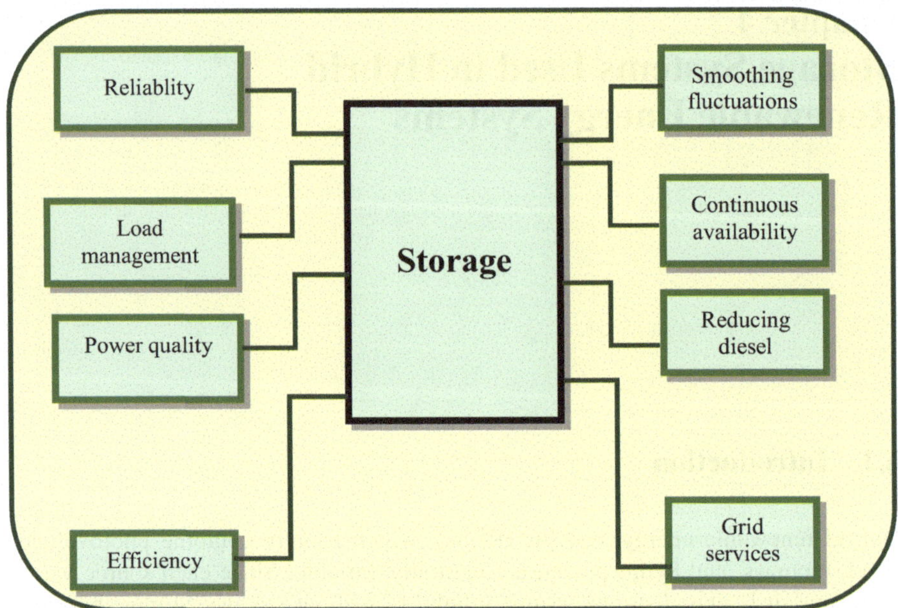

Fig. 4.1 Roles of storage in hybrid renewable energy systems

4.2.2.1 Supply and Demand Balance

In order to stabilize power quality and reduce unused energy, storage systems serve as a link between fluctuating demand and generation. They collect excess electricity when renewable output is high and release it when production declines. For instance, in a PV/wind/battery system, peak generation may surpass the instantaneous load; instead of dumping this excess, the controller transfers it into the battery [11–14].

4.2.2.2 Enhancing Reliability and Continuity of Supply

An ESS maintains power in remote or off-grid locations where the utility grid is nonexistent or unstable. Storage keeps voltage and frequency steady, allowing essential loads like irrigation, communications, lighting, and basic medical equipment run smoothly by storing daytime excess from renewable sources and releasing them when production decreases. Consider an agricultural microgrid. While wind and solar power may be at their highest at noon, some crops require water at night. The ESS uses this excess energy to power the pumps at night, making intermittent generation into reliable service [11–14].

4.2.2.3 Smoothing Renewable Energy Fluctuations

Because wind and solar power are not flexible, they can generate when the climate permits it rather than when the system requests it, their output fluctuates irregularly, causing voltage and frequency deviations. The missing bridge is provided by an energy storage system (ESS), which balances the power flow, and returns energy during short-term deficits when production falls and absorbs short-term excess when generation suddenly exceeds demand. As a result, voltage and frequency are more closely controlled, which is crucial in standalone systems and microgrids without a big grid [11–14].

4.2.2.4 Reducing Diesel Generator Runtime and Fuel Costs

In PV/Wind/Diesel/Battery systems, the battery controls, when and how, the generator is used. When the generator is needed, the battery reduces ramps and covers transients, allowing the generator system to operate in its efficient operation instead of idling (a generator runs at very low load) or short-cycling (frequent start/stop sequences). During low-load hours, stored renewable energy meets the demand, negating the need for the diesel to start at all. Lower fuel usage, lower CO_2 emissions, and fewer maintenance events and longer service intervals are quantifiable results. In summary, storage reduces the consumption of fossil fuels and makes every generator hour significant [15–18].

4.2.2.5 Load Management and Peak Shaving

ESSs are essential to the efficiency of hybrid systems. Storage systems can avoid overloading generators or depending on costly grid imports by using peak shaving, which allows them to release energy during high demand. They can charge during low demand or during periods of low-cost energy and discharge during periods of high demand by using load shifting. When combined, these methods not only guarantee a more dependable power source but also increase the hybrid energy system's overall effectiveness and economic performance. For instance, Fig. 4.2 illustrates how energy storage affects the load profile. High demand peaks and extensive fluctuations are experienced by the system in the absence of storage.

4.2.2.6 Supporting Grid Services and Frequency Regulation

Storage in grid-connected hybrid systems serves not only the local load but also the grid as a whole in addition to keeping the power on. Additionally, they offer reactive power adjustment and voltage support, which maintain a reliable and continuous electrical flow. In addition to providing stability, those characteristics act as supplementary services that enable hybrid systems to protect the grid and increase

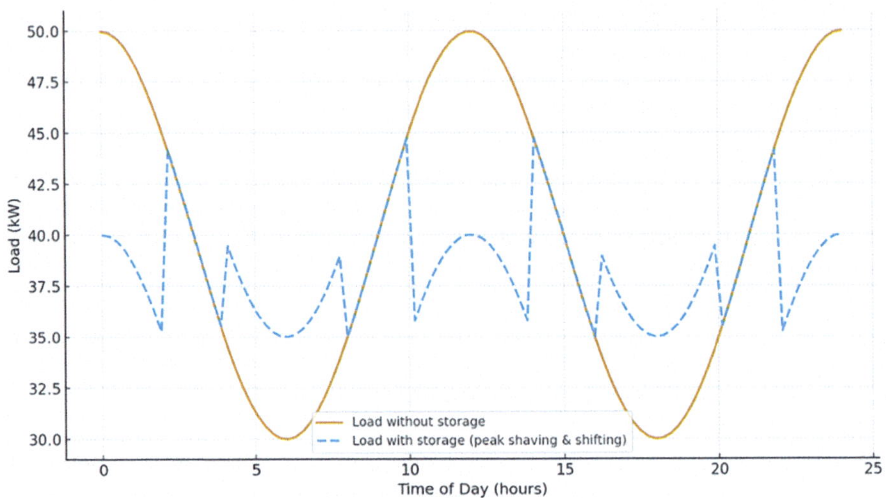

Fig. 4.2 Effect of energy storage on load profile

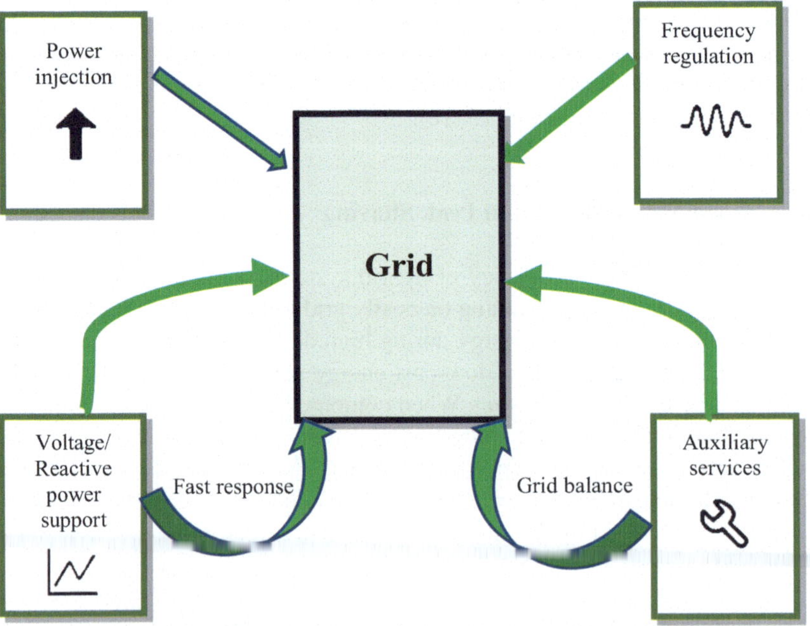

Fig. 4.3 Role of energy storage in supporting grid services within hybrid energy systems

profitability battery [11–13]. Figure 4.3 illustrates how grid services are supported by ESSs in hybrid RESs. The roles of voltage/reactive power support, fast frequency response, and auxiliary services that improve stability and economic value are shown.

4.2.2.7 Improving System Efficiency and Renewable Penetration

When production exceeds demand, a significant renewable energy produced may be lost if storage is not provided. However, this can be resolved by adding storage, which maximizes use of renewable energy by capturing and using each additional kilowatt-hour. Consequently, renewable energy increases in the energy mix, decreasing reliance on fossil fuels and improving sustainability. For example, in an ideal off-grid configuration, a hybrid photovoltaic system with batteries can achieve over 90% renewable energy penetration, making us closer to full energy independence. A simple bar chart in Fig. 4.4 illustrates how storage significantly increases the use of renewable energy, rising from about 55% without storage to over 90% with storage [11].

With optimal storage, numerous studies (such as PV/battery or PV/battery/supercapacitor) demonstrate that renewable penetration in off-grid microgrids can surpass 85–95%. If the system is a multi-storage architecture and incorporates various storage types, it can even approach 100%

4.2.2.8 Power Quality and Fast Transient Response

Supercapacitors and batteries operate as an energy system's dynamic stabilizers [11–14]. They can respond to abrupt voltage fluctuations or surges in milliseconds, preventing damage to delicate devices and ensuring uninterrupted appliance operation (Table 4.1).

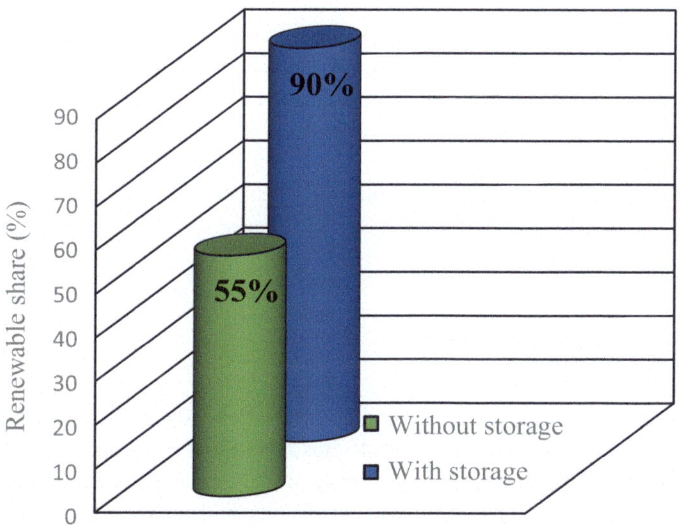

Fig. 4.4 Impact of ESSs on RESs use

Table 4.1 Applications and role of fast response from batteries and SCs

Applications	Role of fast response from batteries & supercapacitors
Hospitals	Prevent life-support machines from shutting down during short power disturbances
Data centers	Keep servers online and protect against data loss when voltage fluctuates
Electric vehicles	Stabilize sudden changes in power demand during fast acceleration or braking

4.2.2.9 Backup Power during Outages

Energy storage increases stability in off-grid and weak-grid hybrid RESs by supplying backup power during interruptions to the system.

4.2.2.10 Flexibility in System Design and Scalability

The flexibility and scalability of hybrid RESs with integrated storage, which enable progressive development and flexibility to fluctuating energy demands, are important benefits. System designers can begin with a small-scale configuration and gradually increase the capacity as load requirements increase or new users are connected thanks to the modular nature of energy storage systems, especially modular battery banks or hybrid storage units (batteries, supercapacitors, or hydrogen systems) [15–18].

Additionally, the flexibility of the system design gives hybridization with a variety of conventional and RE sources, including diesel units, wind turbines, biomass generators, and PV panels, creating a flexible and adaptable microgrid architecture. Under fluctuating load and climatic conditions, this multi-source integration maximizes energy use and improves system reliability.

Energy Management Systems (EMS) are essential for coordinating these many energy sources and storage devices from a control perspective. From home applications to community-level microgrids, advanced EMS systems [11–14], that frequently integrate artificial intelligence (AI), fuzzy logic (FL), or predictive control (MPC), allow for real-time decision-making to balance supply and demand across different levels of control (Fig. 4.5).

4.3 Storage Technologies

In order to balance intermittent generation, satisfy load needs, and preserve system reliability, energy storage technologies, are essential component in hybrid energy systems. Applications can use single storage technologies or multi-storage techniques, which integrate various technologies and make use of their complementing features, depending on the system design. In hybrid energy systems,

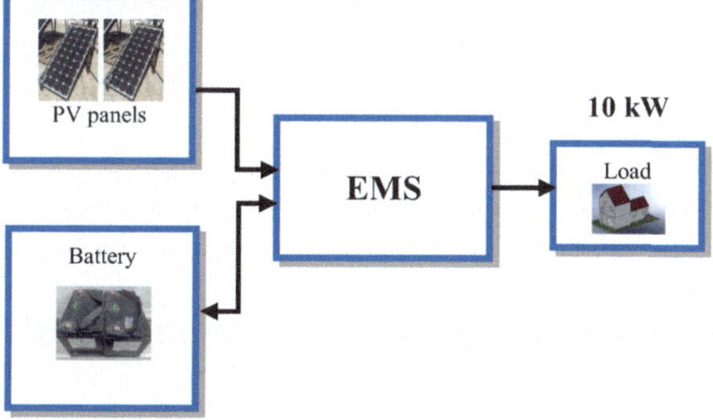

(a) PV with batteries

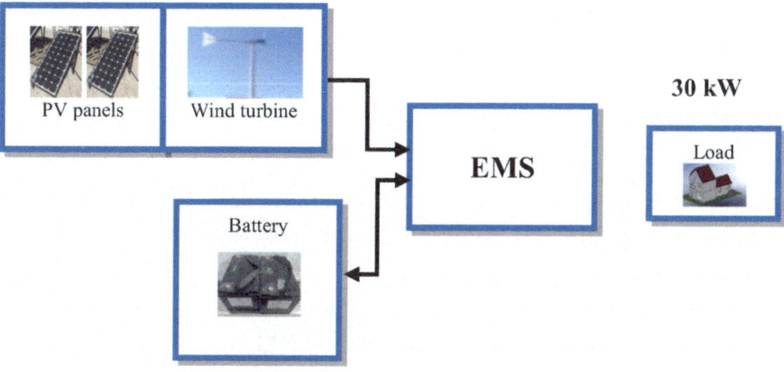

(b) PV/wind turbine with batteries

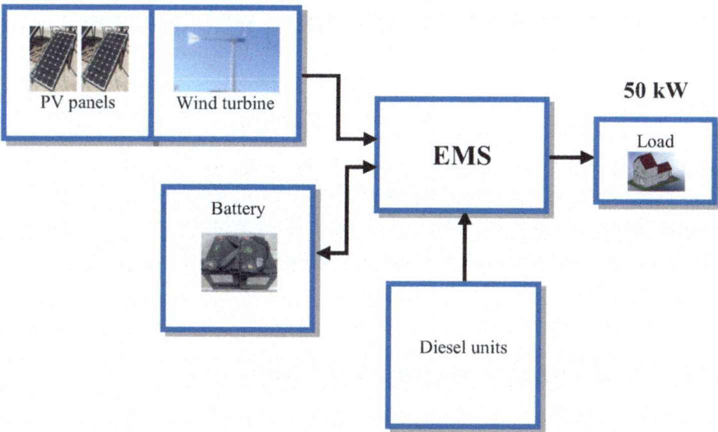

(c) PV/wind turbine with batteries and diesel generator

Fig. 4.5 Hybrid RESs scaling. (**a**) PV with batteries. (**b**) PV/wind turbine with batteries. (**c**) PV/wind turbine with batteries and diesel generator

selecting the appropriate storage type or a mix of storage types is essential. Multi-storage systems are better in terms of performance, resilience, and adaptability, particularly in modern, intelligent, or urgent energy situations, but single storage technologies are simple to use and economical for small-scale or stable applications [16–21].

4.3.1 Single Energy Storage Technologies in Hybrid RESs

The storage needs of hybrid RESs can be satisfied by a single storage technology. Designing, managing, and integrating them is easy, particularly for smaller or simpler systems [19–23] (Table 4.2).

4.3.2 Multi-Energy Storage Systems

Two or more storage technologies are combined in these systems, to take advantage of complementary benefits and compensate for individual limitations. They offer increased efficiency, flexibility, and long-term advantages, especially in complex systems. Table 4.3 below lists the types, applications, advantages, and disadvantages of the main used multi-storage in RESs [24–28].

4.4 Typical Hybrid RES Architectures with Storage

Combining two or more energy sources, most frequently renewables like solar, wind, biomass, and hydro, into a single system is a common practice for hybrid RESs, occasionally with conventional backup like gas turbines or diesel. Smart control systems and storage options (batteries, hydrogen, and pumped hydro) are crucial for maintaining a steady, effective, and dependable energy supply despite fluctuations in the availability of natural resources [15–18] (Fig. 4.6).

4.4.1 PV/Wind Turbine with Battery Storage

Storage is crucial in this system because it balances the inherent unpredictability of both energy sources, providing a consistent power supply even when the quantity of solar or wind energy varies.

Table 4.2 Most popular single energy storage

Technologies	Lithium-ion battery	Lead acid battery	Flywheel	Pumped hydro storage	Supercapacitor	Thermal storage
Common applications	Portable electronics	Automotive starter batteries,	Uninterruptible power supply	Large-scale grid energy storage	Regenerative braking, power backup	Building cooling,
	Electric vehicles	Backup power systems	Frequency regulation	Load balancing	Short-term energy bursts	Industrial process heat storage
	Grid storage	Off-grid solar	Short-term energy storage	RE integration	Frequency regulation	Concentrated solar power plants

Table 4.3 Summary of mainly multi-storage technologies in RESs

Hybrid storage type	Applications
Battery + supercapacitor	Microgrids, Electric vehicles Smartgrids
Battery + flywheel	Wind systems, Critical industrial applications
Battery+thermal storage	Solar systems, Rural agricultural systems, Smart buildings
Battery+pumped hydro	Large-scale PV + hydro systems, Islanded or mountainous grid-tied systems
Li-ion battery+flow battery	Utility-scale hybrid storage, Long-duration+fast-response applications
Supercapacitor+flywheel	High-power applications, transportation, Data centers
Battery+hydrogen (Electrolyzer+fuel cell)	PV systems for long-term seasonal storage Or remote areas
Battery+ CAES	Wind systems for grid-scale flexibility

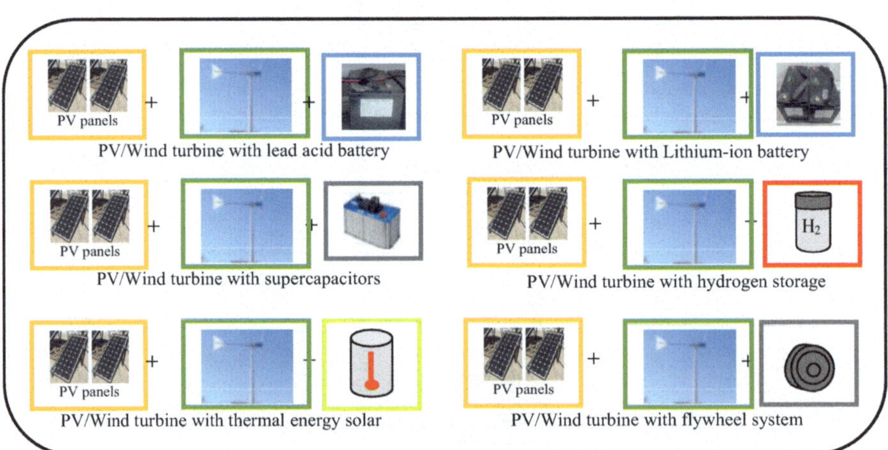

Fig. 4.6 Most important multi sources systems with storage

$$SOC(t+\Delta t) = SOC(t) + \frac{\eta_{charge}.P_{charge}(t) - P_{discharge}(t).\eta_{discharge}}{E_{max}}\Delta t \qquad 4.1$$

The net power delivered to the load or grid must satisfy:

$$P_{load}(t) = P_{gen}(t) + P_{disch}(t) - P_{char}(t) \qquad 4.2$$

4.4.2 PV/Wind with Battery and Diesel Generator

This system is one of the most practical and reliable options for off-grid areas where renewable energy cannot ensure a consistent supply of electricity. In addition to an ESS and a diesel generator (DG), this system effectively combines two continuous renewable energy sources (solar and wind). Coordinating all these variables results in an economical and effective energy supply, even under fluctuating load demand or weather conditions.

The battery storage system acts as an energy bridge to reduce dependency on the DG. The Energy Management System (EMS) continuously tracks the battery's state of charge (SOC), the availability of renewable energy sources, and the demand for power in order to identify the best power-sharing strategy.

When renewable resources arc insufficient, such as at night when wind is low, the DG starts up right away to meet the load requirement and, if necessary, recharges the battery to maintain a sufficient SOC. This ensures constant power delivery and improves system reliability. Additionally, the EMS can reduce CO_2 emissions, fuel consumption, and operating expenses by prioritizing the usage of RE sources and only planning DG operation when necessary.

4.4.3 PV/Wind with Battery and Supercapacitor

A very effective and flexible energy system that combines renewable energy generation with a dual energy storage system is the PV/Wind with Battery and Supercapacitor hybrid configuration. By combining the complementary qualities of supercapacitors and batteries, this combination aims to increase system performance, efficiency, and longevity. It works particularly effectively in situations like microgrids, rural electrification, or electric mobility systems where sudden load changes and renewable power fluctuations happen often. Through efficient energy sharing, the Energy Management System (EMS) ensures quick transient response, steady DC bus voltage, and longer battery life by coordinating power flow between renewable sources and dual storage units.

4.4.4 PV/Wind with CAES and Supercapacitor

The PV/Wind hybrid configuration with CAES and Supercapacitor is advanced multi-storage system designed to increase the dynamic performance and reliability of RESs. It results in a more stable and efficient system that can control the irregular nature of RE sources by combining the long-duration storage capacity of CAES with the fast-response characteristic of supercapacitors.

By compressing air into an above-ground or subterranean tank using surplus renewable energy, the CAES offers large-scale energy storage. During discharge,

CAES is released to power a turbine connected to a generator, producing electrical energy.

In contrast, the supercapacitor quickly absorbs or injects power to balance changes in renewable output, manage short-term fluctuation, and lower DC bus voltage variations.

4.4.5 Solar PV/Wind with Battery and Flywheel

An effective way to increase the reliability and stability of renewable energy systems is to combine a hybrid solar PV/wind system with flywheel and battery storage. This configuration reduces dependence on a single energy source by using wind turbines and solar panels to produce energy from complementing renewable sources. The battery storage system ensures energy availability during times with low solar irradiation or wind speed by providing medium- to long-term energy storage. In addition, the flywheel storage system is used for short-term energy storage, providing rapid response to sudden changes in demand and controlling power fluctuations. This hybrid storage strategy improves power quality, decreases battery stress, boosts system efficiency, and enhances the energy system's lifespan and stability.

4.5 Common Hybrid Configurations with Storage

HES with a single type of energy storage are widely used for simplicity and cost-effectiveness, while when combining two or more types of energy storage, it improves performance, lifespan, and flexibility (Table 4.4).

4.6 Control & Energy Management Systems in Hybrid RESs

For hybrid RESs that integrates renewable sources and storage units to operate as efficiently as possible, effective energy management and control systems (EMS) are essential. Dynamic, adaptive, and predictive energy supervision is made possible by intelligent controllers such digital twin technologies, and artificial intelligence (AI). These advanced methods guarantee effective state-of-charge balancing, charge and discharge optimization, and energy storage component lifespan extension. Intelligent EMS improves the hybrid system's operational reliability and stability by predicting demand or generation developments and continuously monitoring system conditions. Additionally, it is essential for decreasing energy costs, reducing losses, and enhancing system performance. Hybrid RESs can accomplish long-term economic and environmental benefits, optimal energy usage, and sustainable operation through such smart operation and coordination.

Table 4.4 Common configuration of HES with single and multi-storage

Configuration	Applications
PV + diesel + **Lead-acid battery**	Off-grid villages, Emergency power systems
PV + wind + **Lithium-ion battery**	Off-grid farms, Microgrids
Wind turbine + diesel + **Lead-acid battery**	High-wind off-grid communities
PV + hydro + **Li-ion** battery	Mountainous areas
PV + b**attery** + **supercapacitor**	Water pumping systems, EV charging stations
PV + wind turbine+ **battery** + **flywheel**	Industrial microgrids, Microgrids
PV + diesel + **battery** + **supercapacitor**	Remote telecom stations, Backup systems
PV + b**attery** + **thermal storage** (hot water tank)	Agriculture, Greenhouses, Buildings
Wind turbine+ **battery** + **pumped hydro**	Islanded systems, Mountainous regions
PV + b**attery** + **hydrogen (fuel cell)**	Seasonal storage, Off-grid systems
PV + b**attery** + **flow battery**	Hybrid power plants
PV + wind turbine+ **compressed air** + **battery**	Grid-level storage, Desert
PV + wind turbine + **battery** + **supercapacitor**	Agricultural water pumping

4.6.1 SOC-Based Control (State of Charge Monitoring)

This control's goal is to guarantee that batteries are stored within safe operating parameters in order to avoid overcharging or deep discharge. Based on limit values, the EMS turns on or off battery charging and discharging while continuously monitoring the SOC. Longer battery life, system reliability during variations, and flexibility for emergency load delivery are the advantages of this control.

In general terms, we have:

$$SOC_{\min} \leq SOC \leq SOC_{\max} \tag{4.3}$$

Where: SOC_{min} usually around 20–30% to prevent deep discharge, and SOC_{max}: usually around 90–95% to avoid overcharging.

Different control strategies can be proposed based on this logic algorithm:

- If $SOC < SOC_{min}$ stop discharging, prioritize charging.
- If $SOC > SOC_{max}$ stop charging, use surplus energy elsewhere.

4.6.2 Power Sharing Algorithms

Optimizing the distribution of power generation and storage discharge among the available energy sources is the primary goal of EMS in hybrid RESs. Depending on the system architecture and operational objectives, various power-sharing techniques can be used.

- Priority-based power flow approach: where RE sources, are provided, ESSs come next and DGs are employed as backup supplies.
- Proportional load sharing: on the other hand, ensures that each source contributes equally by dividing the power among them according to their rated capacity or instantaneous availability.
- Droop Control: on the other hand, is especially well-suited for decentralized and microgrid applications since it simulates traditional grid behavior by enabling each source to independently modify its power output.

Together, these approaches help to lower fuel consumption, balance component usage, and provide continuous power source switching, all of which improve the sustainability and reliability of hybrid RESs.

4.6.3 Maximum Power Point Tracking (MPPT) Integration

Optimizing energy extraction from wind turbines or PV panels in spite of often fluctuating environmental parameters is the primary objective of MPPT management. Several MPPT procedures can be applied, depending on the source's characteristics. While Perturb and Observe (P&O), Incremental Conductance (INC), and smart techniques like FLC or neural network-based controllers are common for PV systems, strategies like Tip-Speed Ratio (TSR) and Power Signal Feedback (PSF) are commonly employed for wind turbine systems [29–30].

The EMS determines whether to charge the storage unit, supply the load, when it reaches its capacity. To extract the maximum available renewable power, the MPPT controller interacts with the EMS [31–33] (Fig. 4.7). This integrated control system ensures that RE is used as efficiently as possible to maximize the utilization of available resources for sustainable power generation, decrease energy losses, and increase system efficiency.

4.7 Case Studies and Applications

Numerous industries have effectively adopted hybrid energy systems and advanced energy management techniques, showcasing their flexibility in response to regional resources, load variations, and costs. The practical implementation of multi-storage architectures, MPPT-EMS coordination, and optimal power sharing is demonstrated by the following applications.

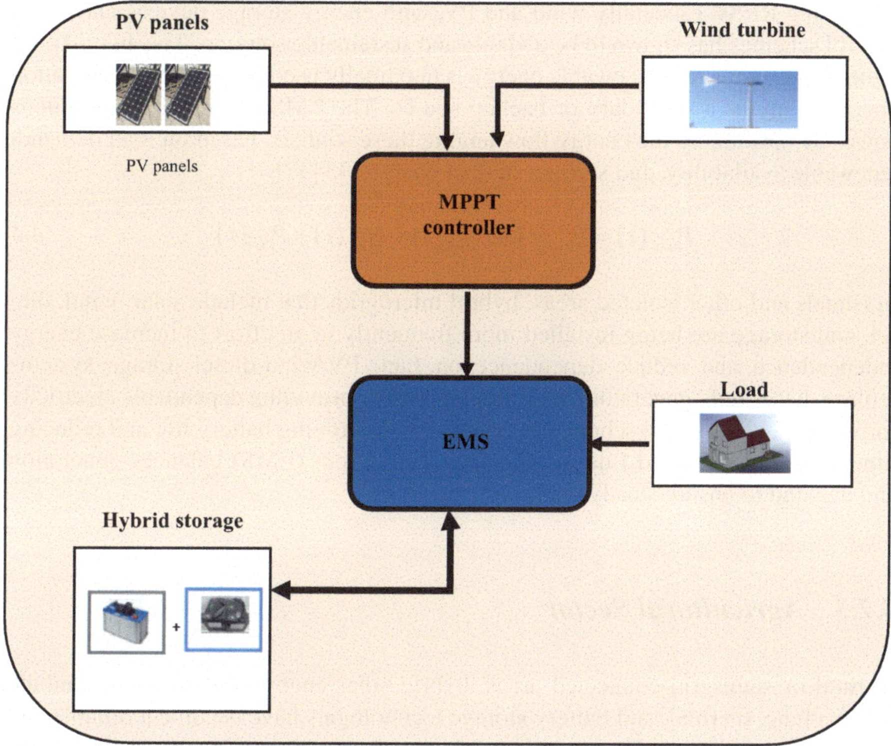

Fig. 4.7 Coordination between MPPT approaches and EMS strategies

4.7.1 Rural Electrification

Solar photovoltaic and battery-based microgrids have become reliable and sustainable energy options in rural areas with limited or non-existent grid connection. PV arrays and lithium-ion or lead-acid battery storage are commonly combined in these systems, which are managed by an EMS that gives priority to the production of renewable energy before discharging the batteries. Using MPPT algorithms, the primary control objective is to optimize solar energy utilization while maintaining battery health through SOC based supervision [34].

4.7.2 Islands and Remote Areas

The lack of electricity access in remote areas and islanded communities frequently makes it challenging to secure a sustainable, economical, and reliable electrical supply. Despite the reliability they provide, diesel generators, the basis of traditional energy systems in these regions, have significant greenhouse gas emissions, high operating costs, and challenges with fuel transportation. For these situations, integrating RESs, especially wind and PV, with energy storage devices and smart control schemes has shown to be a viable and sustainable solution. This hybrid configuration ensures that renewable energy is maximally used, while diesel generation operates only as a secondary or backup source. The EMS maintains a continuous power by optimizing the energy flow among these sources, based on load demand, renewable availability, and storage state of charge [35–37].

$$P_{load}(t) = P_{PV}(t) + P_{wind}(t) + P_{batt}(t) + P_{DG}(t) \quad 4.4$$

In islands and other isolated areas, hybrid microgrids that include solar, wind, diesel, and storage are being installed more frequently in an effort to increase energy independence and reduce dependence on fuel. PV/wind-diesel-storage systems replace diesel-only generators in distant regions by providing dependable electricity for villages, clinics, and schools. In addition to improving battery life and reducing limitation, the advanced Energy Management System (EMS) balances generation and demand to ensure steady operation.

4.7.3 Agricultural Sector

In rural or semi-grid-connected areas, hybrid solar energy systems that combine photovoltaic, thermal, and battery storage technologies have become a reliable and sustainable way to supply energy for agricultural operations. Energy-intensive processes where thermal and electrical needs connect, like crop drying, irrigation, and greenhouse temperature control, are supported by these systems.

The integrated PV-thermal-battery system is designed to minimize dependence on intermittent grid connections or diesel generators while optimizing energy utilization. Usually, it has a PV array that uses solar energy to generate electricity, which is used mostly to supply electrical loads and irrigation pumps. For uses like crop drying or greenhouse heating, a solar thermal collector collects heat energy and stores it in thermal tanks, frequently with the help of water or phase-change materials [36]. To ensure uninterrupted operation during times when solar is not available, the battery bank stores extra electrical energy. By balancing PV generation, battery charge/discharge cycles, and thermal subsystem demands in accordance with load requirements, solar availability, and overall system status, an intelligent EMS coordinates energy flow. For the energy needs of agriculture and rural areas, this coordinated strategy improves system sustainability, reliability, and efficiency

4.7.4 Smart Cities/Microgrids

By effectively managing a variety of energy resources and storage technologies, smart cities integrate microgrids with multi-storage intelligent systems to maximize grid integration. To increase the general effectiveness, robustness, and sustainability of urban energy grids, these multi-energy systems (MES) connect gas, thermal energy, hydrogen, electricity, and various forms of storage. By using advanced predictive analytics and adaptive control techniques, they make it possible to dynamically balance changing demand and renewable generation in real time. These smart multi-storage microgrids, which are controlled by centralized EMS platforms that combine data acquisition and operation control for accurate, optimal performance, encourage zero net energy objectives and facilitate easy integration of RESs and electric vehicles [38].

4.8 Future Directions

Future improvements in hybrid energy storage systems will concentrate on combining innovative developments like solid-state batteries and advanced chemistries with smart hybrid storage technologies like battery-supercapacitor and battery-flow battery combinations. Predictive energy management relies mainly on artificial intelligence (AI), which maximizes the cost-effectiveness and performance of hybrid systems. Blockchain technology improves reliability, safety, and decentralization in energy interactions. The utilization of second-life of electric car batteries, integration of ESSs with the water-energy-food nexus for sustainable uses like aquaculture and desalination, and the requirement for policies that encourage the use of multi-storage hybrid systems are other developments (Fig. 4.8). Together, these developments try to establish energy systems that are more intelligent, reliable, and sustainable [39–44].

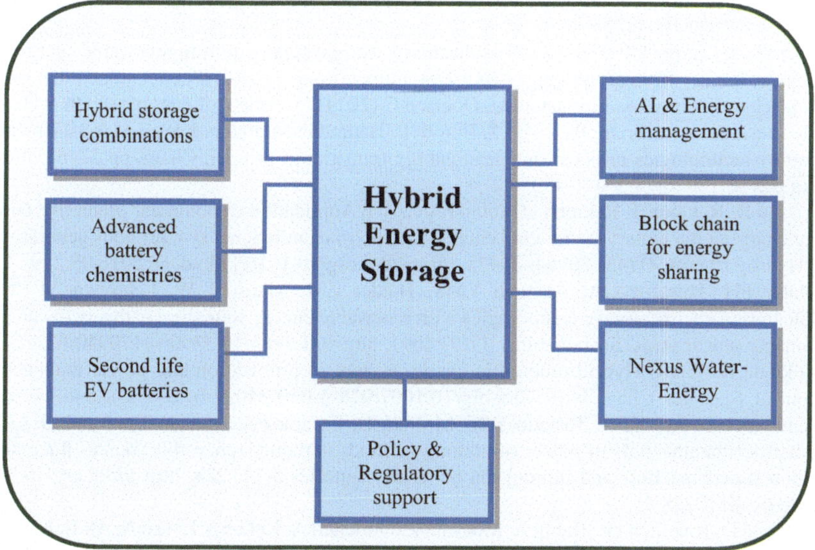

Fig. 4.8 Future direction in hybrid energy storage

4.9 Conclusion

The transition from basic single-storage systems to more complex dual-storage and advanced multi-storage designs is a consequence of the increasing amount of energy demands and the necessity for greater efficiency. The development of advanced EMS is a main factor in the future sustainability of multi-storage hybrid microgrids, particularly in smart cities and isolated grids. These systems effectively coordinate energy flows, optimize economic benefits, and reduce operating costs via real-time data, advanced control algorithms, and analytical forecasting. In addition to increasing the life of storage components, intelligent EMS solutions ensure increased power system stability, resilience, and sustainability.

References

1. Rekioua D (2024) Wind power electric systems: modeling, simulation and control. Springer, Cham. https://doi.org/10.1007/978-3-031-52883-5
2. Tang X, Qi Z, Kong L (2025) The role of energy storage in power systems. In: Electrical energy storage technologies and applications. Springer, Singapore. https://doi.org/10.1007/978-981-96-5625-7_1
3. Liu X, Li W, Guo X, Su B, Guo S, Jing Y, Zhang X (2025) Advancements in energy-storage technologies: a review of current developments and applications. Sustainability 17(18):8316. https://doi.org/10.3390/su17188316
4. Rekioua D, Rekioua T, Soufi Y (2015) Control of a grid connected photovoltaic system. In: 2015 international conference on renewable energy research and applications (ICRERA). IEEE, New York, pp 1382–1387. https://doi.org/10.1109/ICRERA.2015.7418634
5. Rekioua D, Rekioua T, Elsanabary A, Mekhilef S (2023) Power management control of an autonomous photovoltaic/wind turbine/battery system. Energies 16(5):2286. https://doi.org/10.3390/en16052286
6. Castillo A, Gayme DF (2014) Grid-scale energy storage applications in renewable energy integration. Energy Convers Manag 87:885–894. https://doi.org/10.1016/j.enconman.2014.07.063
7. Schischke E, Grevé A, Ehrenstein U, Doetsch C (2024) Overview of energy storage technologies besides batteries. In: Passerini S, Barelli L, Baumann M, Peters J, Weil M (eds) Emerging battery technologies to boost the clean energy transition. Springer, Cham, pp 53–68. https://doi.org/10.1007/978-3-031-48359-2_4
8. Aissou R, Rekioua T, Rekioua D, Tounzi A (2016) Application of nonlinear predictive control for charging the battery using wind energy with permanent magnet synchronous generator. Int J Hydrog Energy 41(45):20964–20973. https://doi.org/10.1016/j.ijhydene.2016.05.249
9. Malik FH, Hussain GA, Alsmadi YM, Haider ZM, Mansoor W, Lehtonen M (2025) Integrating energy storage technologies with renewable energy sources: a pathway toward sustainable power grids. Sustainability 17(9):4097. https://doi.org/10.3390/su17094097
10. Rekioua D (2019) Hybrid renewable energy systems: optimization and power management control. Springer, Cham. https://doi.org/10.1007/978-3-030-34021-6
11. Benavides D, Arévalo P, Tostado-Véliz M, Vera D, Escamez A, Aguado JA, Jurado F (2022) An experimental study of power smoothing methods to reduce renewable sources fluctuations using supercapacitors and lithium-ion batteries. Batteries 8(11):228. https://doi.org/10.3390/batteries8110228
12. Rekioua D, Kakouche K, Babqi A, Mokrani Z, Oubelaid A, Rekioua T, Azil A, Ali E, Alaboudy AHK, Abdelwahab SAM (2023) Optimized power management approach for photovoltaic

systems with hybrid battery-supercapacitor storage. Sustainability 15(19):14066. https://doi.org/10.3390/su151914066

13. Hariraman K, Selvam J, Alagirisamy M, Sivaraju SS (2022) Asymmetrical supercapacitor based solar powered automatic emergency light. In: 2022 8th international conference on smart structures and systems (ICSSS). IEEE, New York, pp 1–5. https://doi.org/10.1109/ICSSS54381.2022.9782231
14. Rekioua D, Rekioua T, Elsanabary A, Mekhilef S (2023) Power management control of an autonomous photovoltaic/wind turbine/battery system. Energies 16:2286. https://doi.org/10.3390/en160522863
15. Rehman S, Natrajan N, Mohandes M, Alhems LM, Himri Y, Allouhi A (2020) Feasibility study of hybrid power systems for remote dwellings in Tamil Nadu, India. IEEE Access 8:143881–143890. https://doi.org/10.1109/ACCESS.2020.3014164
16. Belaid S, Rekioua D, Oubelaid A, Ziane D, Rekioua T (2022) Proposed hybrid power optimization for wind turbine/battery system. Period Polytech Electr Eng Comput Sci 66(1):60–71. https://doi.org/10.3311/PPee.18758
17. Memon ZA, Akbari MA (2023) Optimizing hybrid photovoltaic/battery/diesel microgrids in distribution networks considering uncertainty and reliability. Sustainability 15(18):13499. https://doi.org/10.3390/su151813499
18. Faiz A, Rehman A (2015) Hybrid renewable energy systems: hybridization and advance control. In: 2015 power generation system and renewable energy technologies (PGSRET). IEEE, New York, pp 1–5. https://doi.org/10.1109/PGSRET.2015.7312256
19. Padiyar KR, Kulkarni AM (2019) Solar power generation and energy storage. In: Dynamics and control of electric transmission and microgrids. IEEE, New York, pp 391–414. https://doi.org/10.1002/9781119173410.ch11
20. Boya M, Tekaya K, Spaeth U, Schmuelling B (2025) Exploring the methods and future directions of direct electrical energy storage systems: a systematic review. In: 2025 international conference on clean electrical power (ICCEP). IEEE, New York, pp 930–937. https://doi.org/10.1109/ICCEP65222.2025.11143769
21. Wei P, Abid M, Adun H, Kemena Awoh D, Cai D, Zaini JH, Bamisile O (2023) Progress in energy storage technologies and methods for renewable energy systems application. Appl Sci 13(9):5626. https://doi.org/10.3390/app13095626
22. Carpenter K (2018) Compressed air systems. In: Energy management handbook. River Publishers, Aalborg, pp 583–618
23. Khodadoost Arani AA, Gharehpetian GB, Abedi M (2020) A novel control method based on droop for cooperation of flywheel and battery energy storage systems in islanded microgrids. IEEE Syst J 14(1):1080–1087. https://doi.org/10.1109/JSYST.2019.2911160
24. Berrada A, Emrani A, Achour Y (2026) Optimizing renewable power systems: hybrid gravity-battery energy storage system for wind/PV integration and load balancing. In: Scipioni R, Gil Bardají ME, Barelli L, Baumann M, Passerini S (eds) Hybrid energy storage, vol 47. Springer, pp 43–69. https://doi.org/10.1007/978-3-031-97755-8_3
25. Rekioua D, Mezzai N, Mokrani Z et al (2024) Effective optimal control of a wind turbine system with hybrid energy storage and hybrid MPPT approach. Sci Rep 14:30013. https://doi.org/10.1038/s41598-024-78847-9
26. Rakhshani E, Ruiz JM, Benavides X, Dominguez E (2024) Enhancing low-inertia power systems with grid forming based hybrid energy storage technology. In: 2024 12th international conference on smart grid (icSmartGrid). IEEE, CHAM, pp 464–468. https://doi.org/10.1109/icSmartGrid61824.2024.10578086
27. Wang Y, Ji H, Luo R, Liu B, Wu Y (2025) Energy optimization strategy for wind–solar–storage systems with a storage battery configuration. Mathematics 13(11):1755. https://doi.org/10.3390/math13111755
28. Ramos HM, Coronado-Hernández OE, Besharat M, Carravetta A, Fecarotta O, Pérez-Sánchez M (2025) Energy storage systems in micro-grid of hybrid renewable energy solutions. Technologies 13(11):527. https://doi.org/10.3390/technologies13110527

29. Rekioua D (2020) Power electronics in hybrid renewable energies systems. In: Hybrid renewable energy systems. Springer, Cham, pp 39–77. https://doi.org/10.1007/978-3-030-34021-6_2
30. Rekioua D (2020) MPPT methods in hybrid renewable energy systems. In: Hybrid renewable energy systems. Springer, Cham, pp 79–138. https://doi.org/10.1007/978-3-030-34021-6_3
31. Ndwali PK, Njiri JG, Wanjiru EM (2021) Economic model predictive control of microgrid connected photovoltaic-diesel generator backup energy system considering demand side management. J Electr Eng Technol 16(5):2297–2312. https://doi.org/10.1007/s42835-021-00801-w
32. Rekioua D (2020) Hybrid renewable energy systems overview. In: Hybrid renewable energy systems. Springer, Cham, pp 1–37. https://doi.org/10.1007/978-3-030-34021-6_1
33. Biswas P et al (2025) An extensive and methodical review of smart grids for sustainable energy management—addressing challenges with AI, renewable energy integration and leading-edge technologies. IEEE Access. https://doi.org/10.1109/ACCESS.2025.3537651
34. Kakouche K, Oubelaid A, Mezani S, Rekioua D, Rekioua T (2023) Different control techniques of permanent magnet synchronous motor with fuzzy logic for electric vehicles: analysis, modelling, and comparison. Energies 16(7):3116. https://doi.org/10.3390/en16073116
35. Nair UR, Costa-Castelló R (2020) A model predictive control-based energy management scheme for hybrid storage system in islanded microgrids. IEEE Access 8:97809–97822. https://doi.org/10.1109/ACCESS.2020.2996434
36. Abderrahmane N, Brahmia A, Kerboua A, Kelaiaia R (2025) An optimal sizing methodology for a wind/PV hybrid energy production system for agricultural irrigation in Skikda, Algeria. Appl Sci 15(12):6704. https://doi.org/10.3390/app15126704
37. Rekioua D, Roumila Z, Rekioua T (2008) Étude d'une centrale hybride photovoltaïque–éolien–diesel. J Renew Energ 11(4):623–633. https://doi.org/10.54966/jreen.v11i4.112
38. Martínez Ceseña EA, Mancarella P (2019) Energy systems integration in smart districts: robust optimisation of multi-energy flows in integrated electricity, heat and gas networks. IEEE Trans Smart Grid 10(1):1122–1131. https://doi.org/10.1109/TSG.2018.2828146
39. Akram MN, Abdul-Kader W (2024) Repurposing second-life EV batteries to advance sustainable development: a comprehensive review. Batteries 10(12):452. https://doi.org/10.3390/batteries10120452
40. Taherdoost H (2024) Blockchain integration and its impact on renewable energy. Computers 13(4):107. https://doi.org/10.3390/computers13040107
41. Martino M, Di Rienzo R, Baronti F, Roncellai R, Saletti R (2025) Enhancing second-life battery performance with dynamic equalization: a residential case study. IEEE Open J Indust Electr Soc 6:1864. https://doi.org/10.1109/OJIES.2025.3636758
42. Sutikno T, Arsadiando W, Wangsupphaphol A, Yudhana A, Facta M (2022) A review of recent advances on hybrid energy storage system for solar photovoltaics power generation. IEEE Access 10:42346–42364. https://doi.org/10.1109/ACCESS.2022.3165798
43. Adeyinka AM, Esan OC, Ijaola AO et al (2024) Advancements in hybrid energy storage systems for enhancing renewable energy-to-grid integration. Sustain Energy Res 11:26. https://doi.org/10.1186/s40807-024-00120-4
44. Sayed K, Aref M, Abokhalil G, Mossa MA (2025) Hybrid and advanced energy storage systems: integration, applications, and future trends. IntechOpen, London. https://doi.org/10.5772/intechopen.1010067

Chapter 5
Storage Systems Modeling

5.1 Introduction

5.1.1 Mathematical Formulation

Accurate modeling of these storage systems is essential for their optimal integration into smart grids and hybrid renewable energy setups. At the heart of ESS modeling lies the mathematical formulation of the energy balance equation, which serves as the central structure that enables dynamic and realistic simulations of various storage technologies. This equation quantifies the energy flow, accounting for input, output, storage, and losses, allowing precise prediction of system behavior under changing operating conditions [1–5]. In this chapter, we start with the fundamentals (energy balance and SOC), and then focus on the physics of each technology: power cells (from empirical to CFD-level detail), batteries (electrical/electrochemical/thermal and hybrid models), supercapacitors (from ideal C to transmission-line models), and flywheels (mechanical, electromechanical, and multiphysics). In this chapter, readers will learn how models couple to power electronic interfaces and how to apply them for simulation studies and control design using MATLAB/Simulink. By aligning the modeling approach with the energy balance formulation, the chapter provides the essential tools for understanding, designing, and optimizing ESS to support renewable energy adoption effectively (Fig. 5.1).

D. Rekioua, *Energy Storage for Renewable Energy Systems*, Green Energy and Technology, https://doi.org/10.1007/978-3-032-19589-0_5

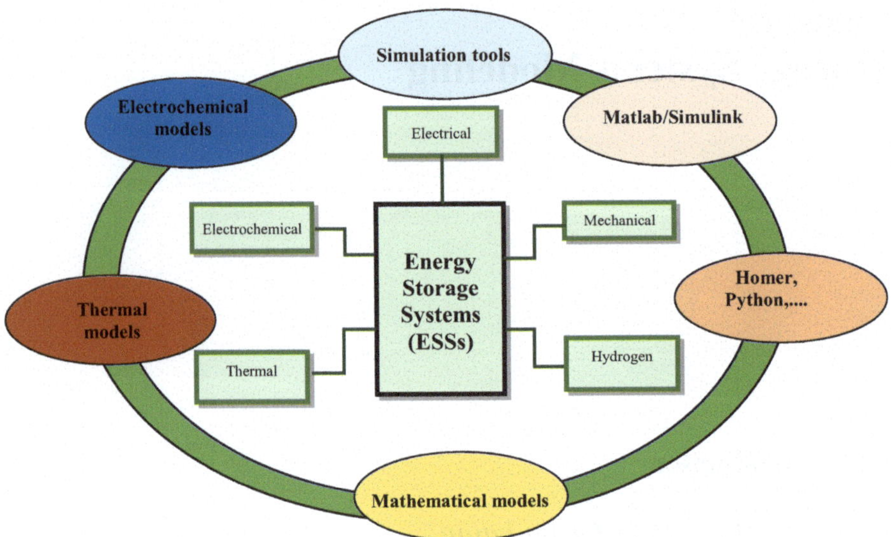

Fig. 5.1 Energy storage systems: types, modeling approaches, and simulation

5.2 General Mathematical Modeling Approach

5.2.1 Energy Balance

The energy equation serves as the foundational starting point in the modeling of energy storage systems (ESSs). Specific models then build on this foundation by incorporating detailed electrochemical, thermal, or mechanical dynamics unique to the storage technology [2–10].

The fundamental energy-balance equations for ESSs are:

$$E_{stored}(t) = E_0 + \int_0^t \left(P_{in}(t) - P_{out}(t)\right) . \eta(t) . dt \tag{5.1}$$

$$P_{BESS}(t) = -K_{droop} \cdot (f - f_{nom}) \tag{5.2}$$

5.2.2 State of Charge Modeling

State of Charge (*SOC*) dynamics describes the evolution of the available energy stored in a battery or energy storage system relative to its full capacity over time. SOC is a crucial parameter that indicates how much charge remains, directly influencing the system's operational decisions, efficiency, and lifespan. The basic SOC dynamics can be modeled by integrating the charge and discharge currents over time:

$$SOC(t) = SOC(t_0) + \frac{1}{C_{nom}} \int_0^t I_{batt}(t) \,.\, \eta_{charge/discharge} \,.\, dt \tag{5.3}$$

$$SOC(t) = \frac{E(t)}{E_{\max}} \tag{5.4}$$

$$\begin{aligned} E(t) &= E(t_0) + \eta \,.\, \int_{t_0}^{t} P(t) \,.\, dt \\ &= E(t_0) + \eta \,.\, \int_{t_0}^{t} V(t) \,.\, I(t) \,.\, dt \end{aligned} \tag{5.5}$$

5.3 Single Storage Modelling

5.3.1 Battery Modelling

Electrical, electrochemical, thermal, and hybrid models are the four primary categories of battery modeling. In order to simulate electrical behaviors like voltage, current, and transient response, electrical models, as Equivalent Circuit Models (ECMs), simulate batteries using circuit components like resistors, capacitors, and voltage sources [10–12]. These models offer a balance of simplicity and accuracy and are commonly used for battery management in electric vehicles and grid storage (Fig. 5.2).

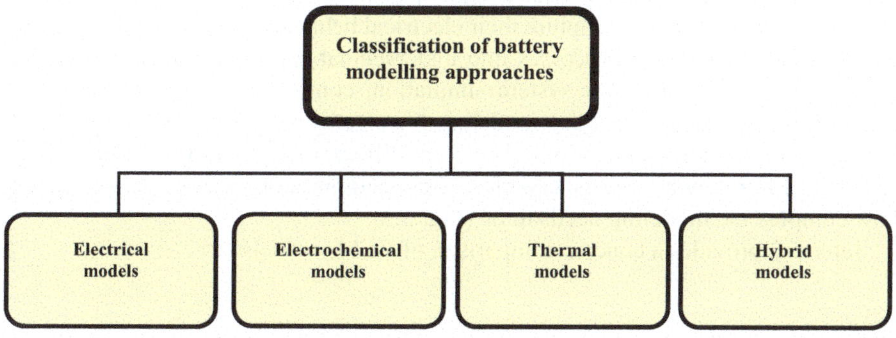

Fig. 5.2 Classification of battery modeling approaches

Table 5.1 Comparative table of main battery model types

Model type	Principle	Accuracy	Applications
Electrical (ECM)	Represent battery with electrical circuit elements (R, C, voltage source).	★★	Energy management, EV powertrain control, microgrids, Real-time BMS
Electrochemical	Describe internal chemical/ physical processes	★★	Research, Battery design, Lifetime prediction, advanced BMS
Thermal	Represent heat generation & dissipation inside the battery.	★★	Thermal management in EVs/ ESS, Safety, cooling system design
Hybrid	Combine ECM + thermal or electrochemical + thermal	★★★★	Accurate SOC/SOH estimation, EVs,

Electrochemical models are more detailed and complex, based on the fundamental chemical reactions and transport phenomena inside the battery, described by partial differential equations. These models offer high accuracy useful for design and diagnostic purposes but demand significant computational resources [13–15].

Thermal models focus on the heat generation and dissipation within batteries, crucial for managing safety and performance under various operating conditions [16, 17].. Lastly, hybrid models combine aspects of electrical, electrochemical, and thermal modeling to capture the coupled interactions between electrical performance, chemical processes, and thermal effects. These integrated models provide a comprehensive set of analytical tools for accurate simulation and control, enhancing battery safety, reliability, and efficiency across applications [18–20].

Table 5.1 shows a comparison of various models in terms of applications, accuracy, and complexity [10, 12].

5.3.1.1 Electrical Equivalent Circuit Models (ECMs)

Electrical ECMs represent batteries using circuit elements such as resistors, capacitors, and voltage sources to capture their electrical behavior. These models simplify complex electrochemical processes into understandable and computationally efficient forms, widely used for system simulation, control, and diagnostics. These ECMs form the basis of battery modeling for electric vehicles, grid storage, and hybrid energy system applications. There are different Electrical ECMs for batteries. They differ in components, equations, applications, and accuracy depending on how complex the modeling needs to be (Fig. 5.3).

Table 5.2 provides a concise comparison of ECMs models.

Table 5.2 Comparison of main ECMs models [13–15]

Model			Components	Equations	Accuracy
Simple models	Ideal voltage source model		Only an ideal voltage source (open-circuit voltage) V_{oc}	$V = V_{oc}$	★/★★
	R_{int} model		V_{oc} Internal resistance R_{int}	$V_{batt} = V_{oc} - I.\ R_{int}$	★★
	Thevenin Models	One RC branch (1st order Thevenin)	V_{oc}, series resistance R_0 One RC parallel network (R_1,C_1)	$V_{batt} = V_{oc} - I.\ R_0 - V_{RC}$ $\frac{dV_{RC}}{dt} = \frac{1}{C_1}\left(1 - \frac{V_{RC}}{R_1}\right)$	★★★
		Higher-order Thevenin (multi RC branches)	Several R_iC_i branches	$V(t) = E_{ocv}.SOC(t) - I(t).R_0 - \sum_{k=1}^{n} V_{RC_k}$ $\frac{dV_{RC_k}(t)}{dt} = -\frac{1}{R_k.C_k}V_{RC_k}(t) + \frac{1}{C_k}I(t)$ $k = 1,2,\ldots\ldots\ldots n$ $\frac{dSOC(dt)}{dt} = -\frac{\eta.I(t)}{C_{nom}}$	★★★★

(continued)

Table 5.2 (continued)

Model		Components	Equations	Accuracy
Semi-advanced model	Dual polarization model (DP)/ Second-order model	V_{oc}, R_0 Two RC networks (R_1,C_1) and R_2,C_2)	$V_{batt} = V_{ocv} - I.R_0 - V_{RC_1} - V_{RC_2}$ $\frac{dV_{RC_1}}{dt} = \frac{1}{C_1}\left(I - \frac{dV_{RC_1}}{R_1}\right)$ $\frac{dV_{RC_2}}{dt} = \frac{1}{C_2}\left(I - \frac{dV_{RC_2}}{R_2}\right)$	★★★★
	Two-RC model	V_{oc} Series internal resistance R_0 Two or more RC parallel branches (R_1,C_1, R_2,C_2, etc.)	$V_{batt} = V_{oc} - I.R_0 - V_{RC_1} - V_{RC_2}$ $\frac{dV_{RC_i}}{dt} = \frac{1}{C_i}\left(I - \frac{dV_{RC_i}}{R_i}\right)$	★★★★

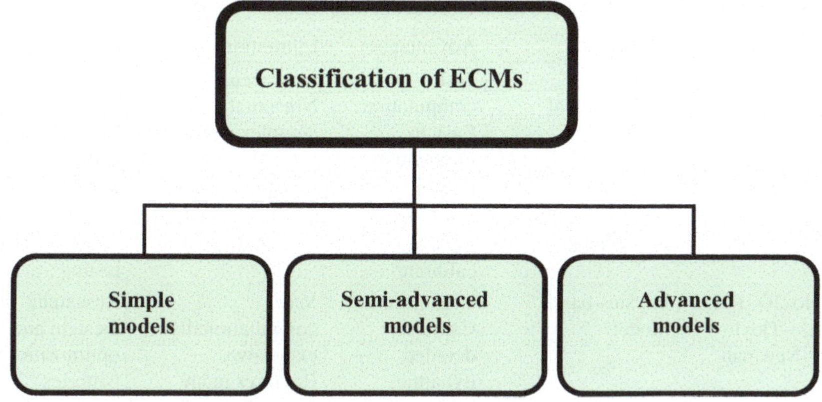

Fig. 5.3 Classification of ECMs

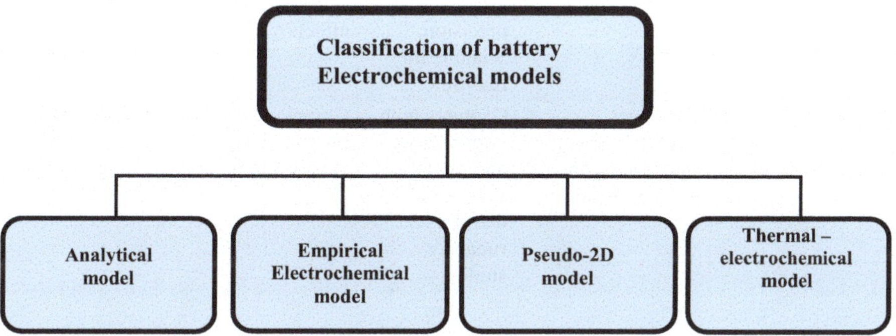

Fig. 5.4 Classification of battery electrochemical models

5.3.1.2 Electrochemical Models

Electrochemical models provide the most physically accurate description, since they represent the internal reactions and transport mechanisms occurring in electrodes and electrolyte. These models range from simple analytical or empirical approaches, which offer rapid estimations but limited accuracy, to advanced physics-based formulations (Fig. 5.4).

The comparison of these different models are represented in Table 5.3.

5.3.1.3 Thermal Models: Thermal Capacity, Heat Generation During Cycling

Thermal modeling is essential to predict temperature distribution, ensure safety, and optimize performance. Several thermal models exist, depending on complexity and accuracy. Figure 5.5 shows a comparative schematic of battery thermal models (From 0D Lumped to 3D CFD approaches).

Table 5.3 Comparative table for the main electrochemical models [13–15]

Model	Description	Advantages	Limitations	Applications
Analytical model	Simplified mathematical equations	Fast computation, Easy to implement	Low accuracy, Not suitable for complex conditions	Quick estimation of battery voltage and behavior
Empirical electrochemical model	Based on experimental fitting of parameters	Simple, Data-driven, Easy to calibrate	Not generalizable across chemistries or conditions	BMS calibration, Aging/lifetime testing
Pseudo-2D (P2D) model—Doyle-fuller-Newman	Physics-based model	High accuracy, Captures detailed dynamics	Very computationally expensive, Requires many parameters	Research, Design, and optimization of batteries
Single particle model (SPM)	Simplified version of P2D	Balance between efficiency and precision, Suitable for real-time	Less accurate under high C-rates or strong electrolyte effects	Control-oriented applications, BMS algorithms
Thermal-Electrochemical Coupled model	Combines electrochemical equations with thermal dynamics	Captures both performance and safety, Useful for thermal runaway studies	Complex, requires thermal data, Heavy computation	EV packs, Safety analysis, Energy storage design

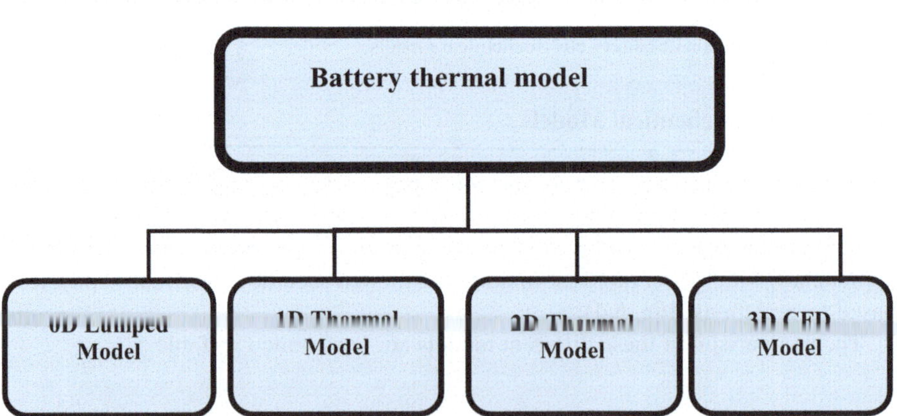

Fig. 5.5 Classification of Battery Thermal Models (0D–3D)

Table 5.4 Comparative table in terms of components, equations, applications and accuracy [16, 17]

Model	Main components	Applications	Accuracy
0D lumped model	Treats battery as a single thermal node	Fast simulations, Battery management systems, Control design	★
1D thermal model	Battery represented along one spatial dimension	Cylindrical cells, Cooling analysis	★★
2D thermal model	Considers two spatial dimensions	Thermal behavior, Module-level studies	★★★
3D CFD model	Full 3D geometry of battery	Pack-level cooling, EVs, Detailed design optimization	★★★★

Table 5.4 is a comparative table of battery thermal models from 0D to 3D, showing the main components, equations, applications, and accuracy.

5.3.1.4 Hybrid models

Several hybrid battery models exist, combining electrical, electrochemical, and thermal domains to capture coupled interactions within the cell. Depending on the level of integration, these models range from simplified electrical–thermal approaches to advanced electrochemical–thermal and fully coupled electrochemical–electrical–thermal frameworks.

Table 5.5 shows the applications and accuracy of these hybrid models.

Table 5.6 illustrates how hybrid models leverage both physics-based understanding and machine learning flexibility to address challenges in battery modeling, including complexity, nonlinearities, aging, and thermal effects.

5.3.2 Fuel Cells Model

Fuel cell models are crucial to energy storage systems because they provide precise performance prediction, control, and optimization. Depending on the required level, different approaches are used (Empirical and equivalent circuit models, Semi-empirical models, electrochemical and CFD-based models, and hybrid and data-driven models). These models are essential for designing hybrid storage systems, optimizing hydrogen utilization, and enhancing the reliability of renewable energy integration [1–3, 5, 21] (Fig. 5.6).

A comparison of FC models in terms of equations, applications, and accuracy is shown in Table 5.7.

Table 5.5 Comparison table of the different hybrid battery models [18–20]

Hybrid model	Combination	Applications	Accuracy
Electrical–thermal	Equivalent Circuit Model (ECM) + Lumped thermal	EV BMS, Pack cooling design, SOC/temperature estimation	★★★
Electrochemical–thermal	Electrochemical model (Pseudo-2D, Newman) + Thermal PDEs	Safety studies, Fast charging, Thermal runaway prediction	★★★★
Electrical–electrochemical	ECM (fast dynamics) + Reduced electrochemical sub-models	Real-time SOC/SOH estimation, BMS with improved accuracy	★★★/★★★★
Electrochemical–electrical–thermal (full hybrid)	Electrical (ECM) + Electrochemical PDEs + Thermal PDEs	EVs, Aerospace, Safety-critical systems, Energy storage optimization	★★★★★
Data-driven hybrid (physics + ML)	Physics-based model (ECM or electrochemical + thermal) + Machine learning	Digital twins, SOH/SOC prediction, Aging forecasting, Fast onboard estimation	★★★★/★★★★★

Table 5.6 Other Hybrid physics–data-driven models [18–20]

Model type	Applications	Accuracy
Electrochemical + Machine learning (ML)	Soc/soh estimation, Degradation prediction, Light weight surrogate modeling	★★★★
Equivalent circuit + Data-driven layer	Real-time battery state estimation, Adaptive control, Voltage prediction	★★★/★★★★
Thermal model + ML prediction	Battery thermal management, Safety assurance, Fault detection	★★★★
Electrochemical–thermal + Data-driven degradation	Life span prediction, maintenance scheduling, System optimization	★★★★★
Physics-informed neural networks (PINNs)	Surrogate modeling, multiphysics battery simulations, Inverse parameter identification	★★★★★

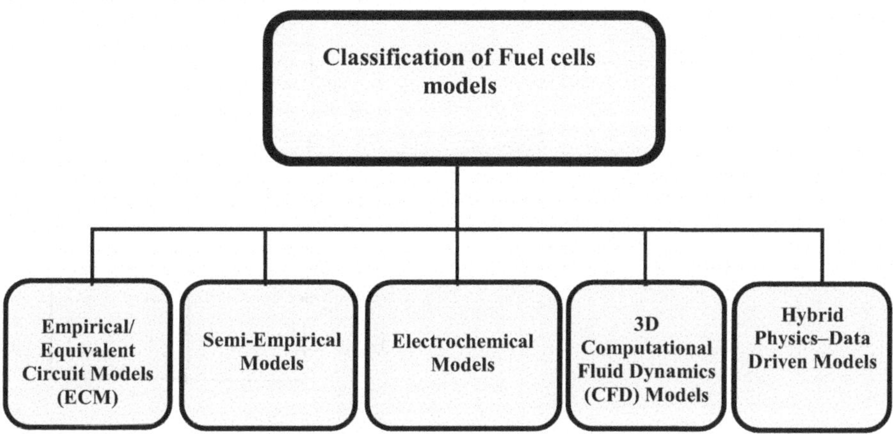

Fig. 5.6 Classification of fuel cell models

Table 5.7 FC models comparison

Model type	Approach	Applications	Accuracy
Empirical/ Equivalent circuit models (ECM)	Electrical + Resistors, Capacitors, Voltage sources	Real-time control, System-level integration, Quick simulations	★★
Semi-Empirical Models	Electrical + Simplified Electrochemical relations	Performance prediction, System optimization, design studies	★★★
Electrochemical Models	Physics-based: Charge, mass transport + Kinetics	Component design, Degradation studies, Utilization analysis	★★★★
3D computational fluid dynamics (CFD) models	Thermal-fluid + Electrochemical	Detailed flow field, Water/heat management, Scaling	★★★★★
Hybrid physics–data driven models	Physics-based + Data-driven	SOH prediction, Fault detection, Daptive control, digital twins	★★★★★

5.3.2.1 Electrical Representation of a Proton Exchange Membrane Fuel Cell (PEMFC)

An equivalent circuit is commonly used to depict the electrical representation of a PEMFC. The electrical representation is as follows (Fig. 5.7).

We have [1]:

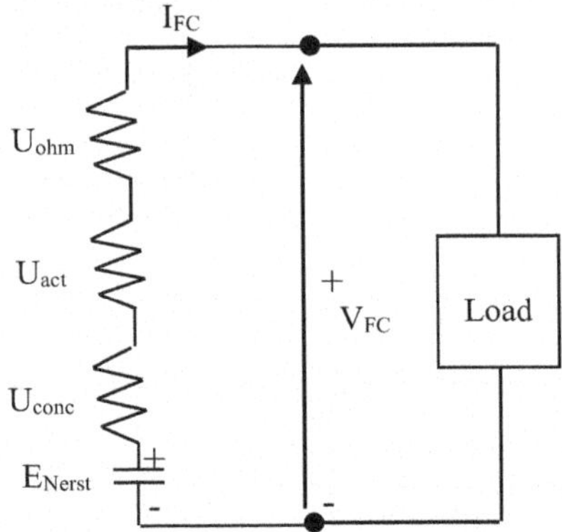

Fig. 5.7 Electrical representation of a PEMFC

$$\begin{aligned} V_{FC}\left(I_{FC},T_{FC},P_{H_2},P_{O_2}\right) &= E_{Nerst}\left(T_{FC},P_{H_2},P_{O_2}\right) \\ &+U_{act}\left(I_{FC},T_{FC}\right)-U_{Ohm}\left(I_{FC},T_{FC}\right)-U_{conc}\left(I_{FC}\right) \end{aligned} \tag{5.6}$$

Where:

$$\begin{aligned} E_{Nerst} &= E_0\left(T_{FC}\right)+\frac{R\,.\,T_{FC}}{2.F}\left(\frac{\ln P_{O_2}^{1/2}\,.\,\ln P_{H_2}}{\ln P_{H_2O}}\right) \\ &= E_0\left(T_{FC}\right)+\frac{R\,.\,T_{FC}}{2.F}\ln\left(\frac{\sqrt{P_{O_2}}\,.\,P_{H_2}}{P_{H_2O}}\right) \end{aligned} \tag{5.7}$$

$$E_0\left(T_{FC}\right)=1.229-8.5\times10^{-4}\left(T_{FC}-298.15\right) \tag{5.8}$$

We will obtain:

$$\begin{aligned} E_{Nerst} &= 1.229-8.5\times10^{-4}\left(T_{FC}-298.15\right)+\frac{R.T_{FC}}{2.F}\left(\frac{\ln P_{O_2}^{1/2}\,.\,\ln P_{H_2}}{\ln P_{H_2O}}\right) \\ &= 1.229-8.5\times10^{-4}\left(T_{FC}-298.15\right)+\frac{R.T_{FC}}{2.F}\ln\left(\frac{\sqrt{P_{O_2}}\,.\,P_{H_2}}{P_{H_2O}}\right) \end{aligned} \tag{5.9}$$

$$E_{Nerst}=\alpha_{n1}+\alpha_{n2}\left(T_{FC}-298.15\right)+\alpha_{n3}\,.\,T_{FC}\left(0.5\ln P_{O_2}+\ln P_{H_2}\right) \tag{5.10}$$

The activation overvoltage U_{act} is:

$$U_{act}=\beta_1+\beta_2\,.\,T_{FC}+\beta_3\,.\,T_{FC}\,.\,\ln(j)+\beta_4\,.\,T_{FC}\,.\,\ln\left(C_{O_2}\right) \tag{5.11}$$

$$C_{O_2}=\frac{P_{O_2}}{5.08\times10^{6}\,.\,\exp\left(-\dfrac{498}{T_{FC}}\right)} \tag{5.12}$$

Also, the Tafel approximation is commonly used:

$$U_{act} \approx \frac{R \,.\, T_{FC}}{\alpha \,.\, F} \ln\left(\frac{I}{I_0}\right) \tag{5.13}$$

The ohmic overvoltage U_{ohm} is:

$$\begin{aligned} U_{Ohm} &= j \times R = j(R_{m,\,area} + R_{c,\,area}) = \frac{I_{FC}}{A_{cell}}\left(\frac{l_m}{\sigma_m(T,\lambda)} + A_{cell} \,.\, R_c\right) \\ &= \frac{I_{FC}}{A_{cell}} \,.\, R_{m,\,area} + I_{FC} \,.\, R_{c,\,area} \end{aligned} \tag{5.14}$$

$$U_{Ohm} = \frac{I_{FC}}{A_{FC}} \cdot \left(\frac{181.6\left[1 + 0.03j + 0.06\left(T_{FC}/303\right)^2 \times J^{2.5}\right]}{\left[\lambda - 0.634 - 3j\right] \exp\left[4.18\left(\frac{T_{FC} - 303}{T_{FC}}\right)\right]} \right) + A_{cell} \,.\, R_c \tag{5.15}$$

The concentration overvoltage U_{conc} is given as:

$$U_{conc} \approx -\frac{R \,.\, T_{FC}}{n.F} \,.\, \ln\left(1 - \frac{I_{FC}}{I_m}\right) \tag{5.16}$$

For small deviations a linearized or empirical form is used:

$$U_{conc} \approx -\beta \,.\, \ln\left(1 - \frac{I_{FC}}{I_{\lim}}\right) \tag{5.17}$$

For a Stack (N cells in series), we will have:

$$\begin{cases} V_{stack} = N \,.\, V_{FC} \\ P_{stack} = V_{stack} \,.\, I_{FC} \end{cases} \tag{5.18}$$

In general, the compact algebraic form of voltage FC is:

$$V_{FC} = E_{Nerst}\left(T_{FC}, P_{H_2}, P_{O_2}\right) - \frac{R \,.\, T_{FC}}{nf} \ln\left(1 - \frac{I_{FC}}{I_{FC-\max}}\right) - j \,.\, R - \frac{R \,.\, T_{FC}}{n \,.\, F} \ln\left(1 - \frac{I_{FC}}{I_{\lim}}\right) \tag{5.19}$$

The production of the hydrogen via the electrolyser is illustrated in Fig. 5.8.
The electrolyser's electrical diagram is displayed in Fig. 5.9.

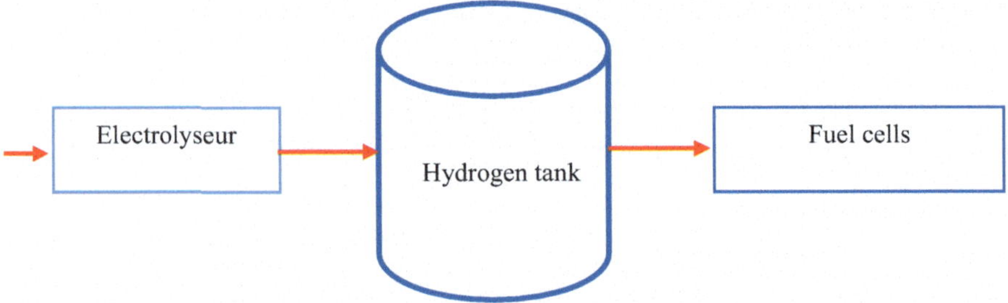

Fig. 5.8 Production of hydrogen

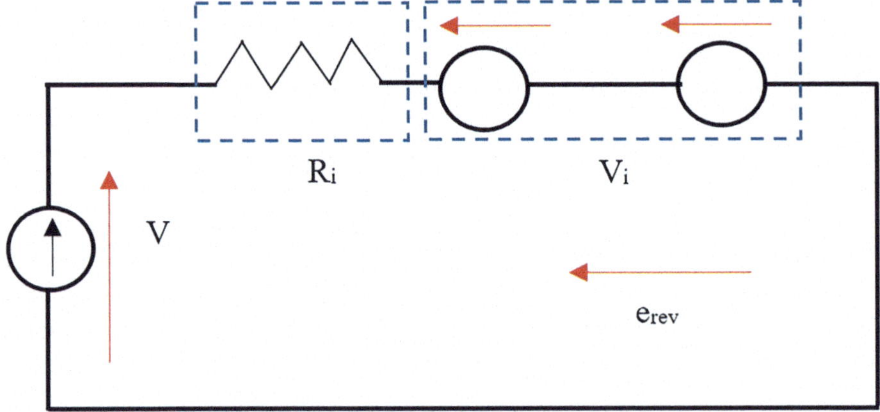

Fig. 5.9 Electrolyser circuit

We model the electrolysis process using the following equation to get the current (*I*), voltage (*V*), and hydrogen characteristics [6–8]:

$$V = \left(\frac{1}{3.064}\right) * I + 1.476 = 0.326I + 1.476 = IR_i + e_{rev} \tag{5.20}$$

The ideal potential V_i is:

$$V_i = \frac{\Delta G}{2F} \tag{5.21}$$

$$\Delta G = 285840 - 163.2\left(273 + T_{\circ C}\right) \tag{5.22}$$

The molar volume V_m given by the following equation:

$$V_m = \frac{R\left(273 + T_{\circ C}\right)}{P} \tag{5.23}$$

The hydrogen production rate which is V_{H2} (ml/min) relative to the input current I(A) is determined by:

$$V_{H2} = V_m(I)\left(\frac{10^3 ml}{l}\right)\left(\frac{60s}{\min}\right)\left(\frac{I\left(\frac{c}{s}\right)}{2F(C)}\right) \tag{5.24}$$
$$= V_m * 10^3 * 60 * \frac{I}{2F}$$

The electrochemical energy of hydrogen per second, P_{H2} is calculated by the following equation:

$$P_{H2} = V_m * 10^3 * 60 * \frac{I}{2F}\frac{2FV_i}{V_m * 10^3 * 60} = V_i * I \tag{5.25}$$

The useful power of the electrolyser is determined by the input current I and the ideal voltage V_i. The equations above indicate this relationship.

The electrolyser's input electrical power P is:

$$P = V_i = I^2 * R_i + I * e_{rev}$$
$$= \left(V_H \frac{2F}{V_m * 10^3 * 60}\right)^2 * R_i + \left(V_H \frac{2F}{V_m * 10^3 * 60}\right) * e_{rev} \tag{5.26}$$

Mathematical equations have been developed to model the storage of gaseous hydrogen produced by the electrolyzer.

$$P_{pres} - P_{pres-i} = z * \frac{N_{H2} * R_{gas} * T_b}{M_{H2} * V_b} \tag{5.27}$$

$$z = \frac{P * V_m}{R_{gas} * T_{^\circ C}} \tag{5.28}$$

5.3.2.2 Application of PV System with Batteries/FCs (See Chap. 8, Sect. 8.3.1)

5.3.2.2.1 Supercapacitors

Supercapacitors (SCs), also called ultracapacitors or electrochemical capacitors, are increasingly used in hybrid ESSs because of their extended cycle life, high power density, and quick charge-discharge capability.

5.3.2.3 Most Used SCs Model

Table 5.8 introduces a comparison of the most used SCs [9–11, 22].

5.3.2.4 RC Branch Model

Figure 5.10 depicts the SC model and we have the various equations [10].

$$V_{sc} = V_c + R_s I_{sc} \tag{5.29}$$

$$V_c = \frac{1}{C}\int I_{sc}\,dt \tag{5.30}$$

5.3.2.5 Modelling with the Two RC Branch Equivalent Circuit

Figure 5.11 depicts the model with two RC branches [9–11, 22].

The top node in the SC model indicates the supercapacitor terminal voltage, V_{sc}. Three parallel branches appear from this node and connect ground:

Table 5.8 Supercapacitor model comparison

Model type	Applications	Accuracy
Ideal capacitor model	Quick estimation, Basic simulations, Preliminary sizing, Basic theoretical studies	★
RC equivalent circuit (ECM)	Hybrid storage systems, microgrids, EVs, Power smoothing	★★
Transmission line model (TLM)	High-frequency analysis, Dynamic response studies, Research simulations	★★★★
Electrochemical model	Detailed electrochemical studies, Design optimization, Performance analysis and research	★★★★
Hybrid model	High fidelity simulations, R&D	★★★★
Constant capacitance model	Basic simulations, Early design stages, Simple control systems	★
Voltage-dependent capacitance model	Advanced simulations, Performance prediction, Design optimization	★★★★
Leakage & aging models	Long-term performance, Maintenance planning	★★★★

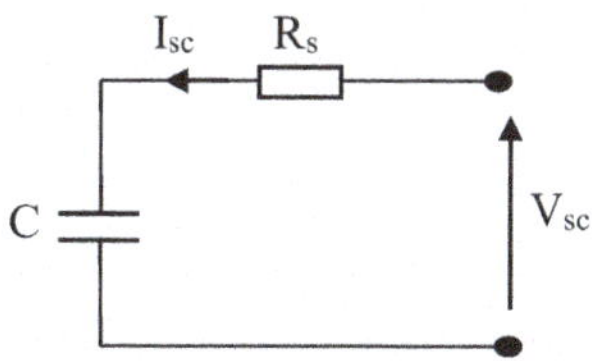

Fig. 5.10 Supercapacitor model

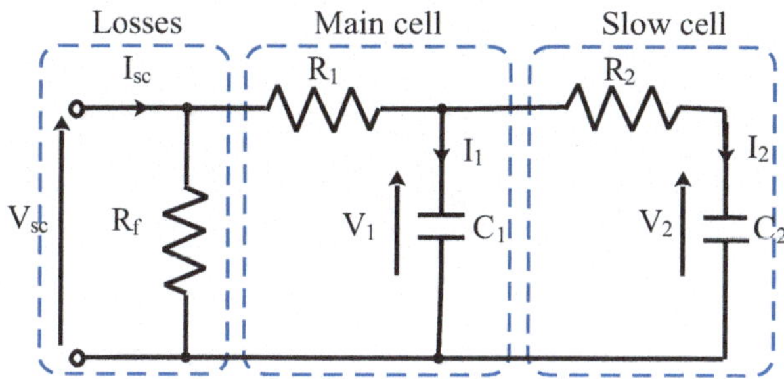

Fig. 5.11 Equivalent electric circuit

- The slow branch, which has a series resistor R_2 and capacitor C_2, with branch current $i_2(t)$ and voltage $V_2(t)$,
- The loss branch, which has a leakage resistor R_f representing a small leakage current,
- The main fast branch, which is composed of a series resistor R_1 and capacitor $C_1$with current $i_1(t)$ and capacitor voltage $V_1(t)$

To complete the current flows in this similar circuit description, the load current i_{load} is indicated as out of the top node, while the external terminal *current* $i_{sc}(t)$ is injected into the top node.

The different equations are:

$$\begin{cases} U_{sc}(t) = N_{sc-s} \,.\, V_{sc}(t) \\ U_{sc}(t) = R_f.(i_{sc}(t) - i_1(t) - i_2(t) - i_{load}(t) \end{cases} \tag{5.31}$$

The terminal node's current balance is provided as follows:

$$i_{sc}(t) = i_f(t) + i_1(t) + i_{12}(t) + i_{load}(t) \tag{5.32}$$

In the slow branch:

$$\begin{cases} \mathrm{i}_2(\mathrm{t}) = C_2 \dfrac{dV_2}{dt} \\ \mathrm{i}_2(\mathrm{t}) = \dfrac{U_{SC} - V_2}{R_2} \end{cases} \tag{5.33}$$

$$C_2 \frac{dV_2}{dt} = \frac{U_{SC} - V_2}{R_2} \tag{5.34}$$

In the main branch:

$$\begin{cases} i_1(t) = i_{sc}(t) - i_2(t) \\ i_1(t) = C_1 \dfrac{dV_1}{dt} \end{cases} \tag{5.35}$$

$$C_1 \frac{dV_1}{dt} = \frac{U_{SC} - V_1}{R_1} \tag{5.36}$$

The leakage branch current is:

$$i_f = \frac{U_{SC}}{R_f} \tag{5.37}$$

The normalized SOC can be expressed as:

$$SOC = \frac{E_{pack} - E_{min}}{E_{max} - E_{min}} \tag{5.38}$$

Where:

$$E_{max} = \frac{1}{2} . C_{pack} \cdot {}^{2}_{pack} \tag{5.39}$$

And E_{min} is the minimum safe energy corresponding to V_{min}.

$$V_{min} = \sqrt{\frac{2 . E_{min}}{C_{pack}}} \tag{5.40}$$

5.3.2.6 Application of PV/Wind Turbine System with SC Storage (See Chap. 8, Sect. 8.3.2)

5.3.2.6.1 Flywheel Modelling

Various modeling techniques are used for flywheel energy storage systems (FESS), depending on the level of application that is needed. They vary into three general categories: integrated, mechanical, and electrical (Fig. 5.12). Other approaches to flywheel modeling can also be divided into three complexity levels. The Ideal mechanical model and the lossy mechanical model are examples of simple models

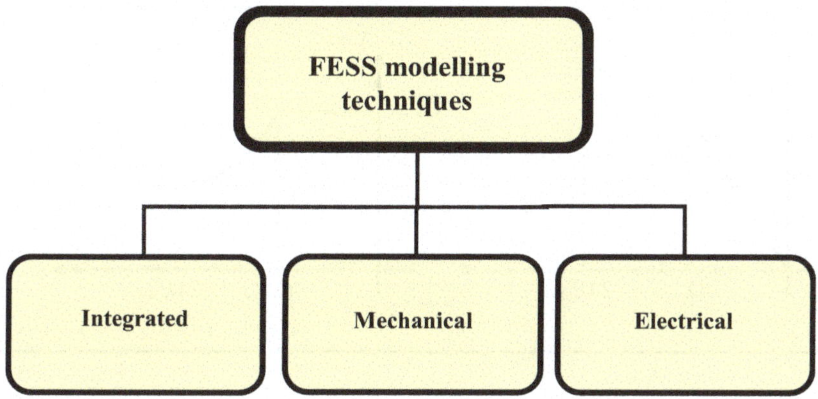

Fig. 5.12 Flywheel models

that are mainly used for loss evaluation and elementary energy estimate. By including the electrical machine and power electronic converter, intermediate models are able to capture more realistic dynamics, which makes them suitable for renewable energy and grid integration applications [13–15]. Lastly, advanced models provide the precise simulations needed for aerospace and high-speed flywheel systems by combining electro-mechanical-thermal connections with nonlinear phenomena.

Other recognized categories of fly wheel models are also used in the analysis and design of FESSs.

5.3.2.7 Application of FESS in Wind Turbine Systems

We make an application using flywheel storage in WTbs. The amount of energy stored is [6]:

$$E = \frac{1}{2} J \, . \, \omega^2 \tag{5.41}$$

Thus the flywheel's reference speed is calculated by:

$$\omega_{ref} = \sqrt{\frac{2 \, . \, E_{ref}}{J_{Tb}}} \tag{5.42}$$

It is evident that (Fig. 5.13):

- For $0 \leq \omega \leq \omega_{rated}$ the torque may be at its maximum, reducing power in proportion to speed.
- For $\omega > \omega_{rated}$, when the power is at its highest and equals the machine's rated power, the electromagnetic torque is inversely proportional to the speed.

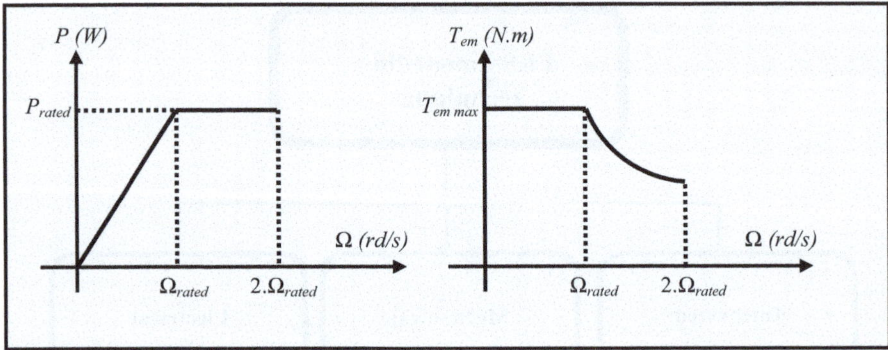

Fig. 5.13 Power and torque evolution

The reference flux is determined by:

$$\phi_{ref} = \begin{cases} \phi_{rated} \;\; if \;\; |\omega| \leq \omega_{rated} \\ \phi_{rated} \cdot \dfrac{\omega_{rated}}{|\omega|} \;\; if \;\; |\omega| > \omega_{rated} \end{cases} \tag{5.43}$$

The whole system's power evaluation is provided by Fig. 5.14:

$$P_{ref} = P_{load} - P_{wind} - \Delta P \tag{5.44}$$

5.3.2.8 Application Under Matlab/Simulink

To control the bus voltage, we need the necessary power. The DC bus voltage is maintained at $V_{DC\text{-}ref}$ = 465 V. The wind power profile employed in the Matlab/Simulink application always supplies the power needed by the load. The evolution of wind power, which ranges from 1000 to 2400 W, is depicted in Fig. 5.15 [6]. It has been observed that the FESS keeps the load's power supply constant. Figure 5.16 shows the outcomes of the other simulations. When wind power exceeds load, the flywheel charges, and when wind power falls short of load, it discharges. This validates the sign convention and expected controlling operation. The rotating speed increases during charge times and decreases during discharge, the slope variations are in agreement with the net power flow and the applied electromagnetic torque. Also, as predicted, the electromagnetic torque responses to changes in speed and power. The DC bus's regulation around $V_{DC,\ ref} \approx$ 465 V, shows that the current/torque loop and the voltage loop operate correctly. During power fluctuations, the stator flux trajectory keeps limited and smooth, suggesting that there are no saturation problems or flux-reducing anomalies [16, 23].

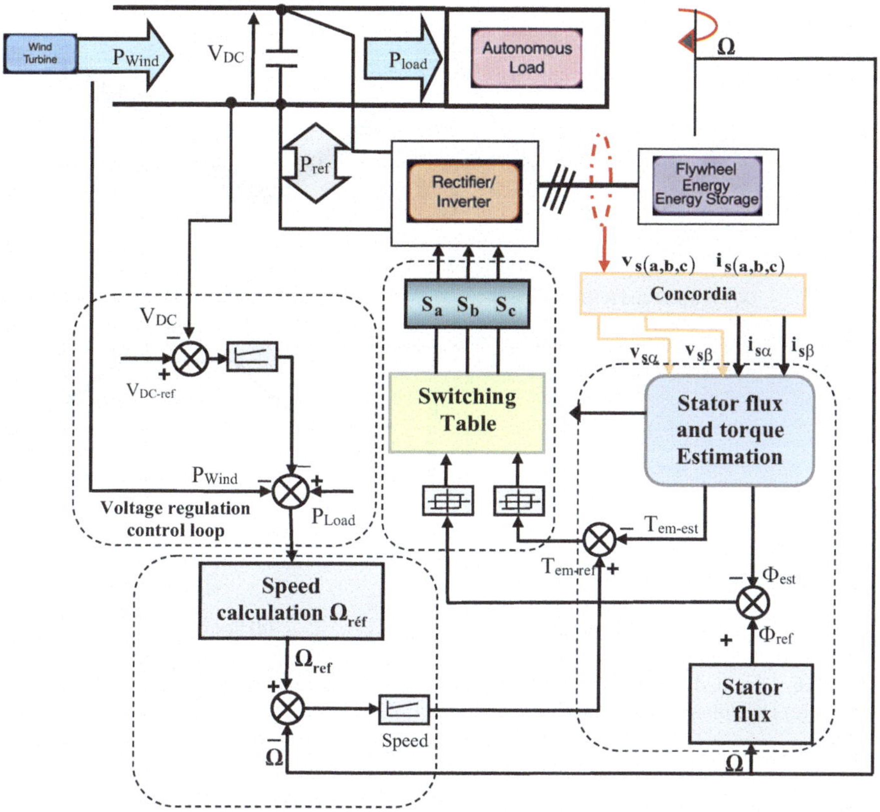

Fig. 5.14 Wind turbine system based on IG with FESS

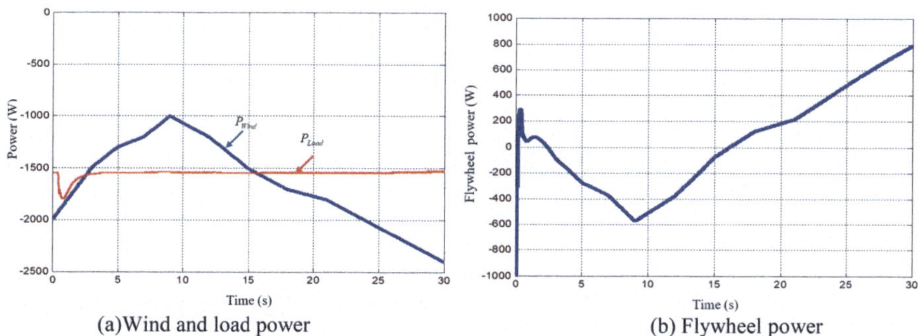

Fig. 5.15 Power waveforms. (**a**) Wind and load power. (**b**) Flywheel power

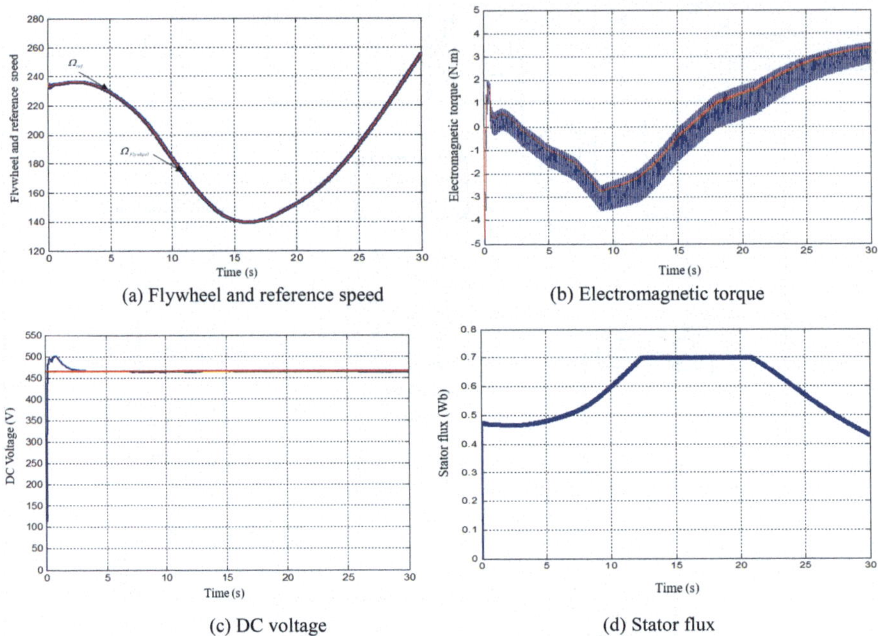

(a) Flywheel and reference speed (b) Electromagnetic torque

(c) DC voltage (d) Stator flux

Fig. 5.16 Obtained simulation results. (**a**) Flywheel and reference speed. (**b**) Electromagnetic torque. (**c**) DC voltage. (**d**) Stator flux

5.4 Dual Storage in Standalone PV Systems

5.4.1 Mathematical Formulation

The mathematical model of dual-storage (Battery + Supercapacitor) for a standalone PV system is given with the following equations. It includes variables and constraints for state of charge, charge/discharge dynamics, efficiency, and power balance [19, 20, 24].

$$P_{load} = P_{PV} + P_{batt} + P_{SC} \tag{5.45}$$

With:

$$\begin{aligned} SOC_{batt}(t+1) &= SOC_{batt}(t) + \frac{\eta_{batt}^{char} \cdot P_{load} \cdot \Delta t}{E_{batt}} - \frac{P_{batt}^{disc} \cdot P_{load} \cdot \Delta t}{\eta_{batt}^{disc} \cdot E_{batt}} \\ SOC_{SC}(t+1) &= SOC_{SC}(t) + \frac{\eta_{SC}^{char} \cdot P_{load} \cdot \Delta t}{SC} - \frac{P_{SC}^{disc} \cdot P_{load} \cdot \Delta t}{\eta_{SC}^{disc} \cdot E_{SC}} \end{aligned} \tag{5.46}$$

$$SOC_{batt,mni} \leq SOC_{batt} \leq SOC_{batt,\max} \tag{5.47}$$

$$SOC_{SC,\min} \le SOC_{SC} \le SOC_{SC,\max} \tag{5.48}$$

5.4.2 Application of PEMFC/SCs Suppling Electric Vehicle (See Chap. 8 Sect. 8.3.3)

Fuel cells and supercapacitors are used for dual storage in an electric vehicle. The two DC/DC converters provide the energy flows. The DC/DC boost converter maintains the voltage at a constant value needed by the electrical load and permits just one direction of energy flow, from the PEMFC to the DC bus. The DC/DC buck-boost converter provides energy in both directions: from the SC to the DC bus and from the DC bus to the SC [25–27].

Determining the references of the current in each source in relation to the electrical load's state (acceleration or braking) and the PEMFC's limitations forms the basis of the control strategy (Fig. 5.17). A bidirectional DC/DC converter connects the SC bank to the PEMFC, whereas a unidirectional DC/DC converter connects the PEMFC to the DC bus [28–33].

The following equations can be used:

$$\begin{cases} C_{dc}\dfrac{dV_{DC}}{dt} = \left(1-d_{fc}\right)i_{fc} + \left(1-d_{sc}\right)i_{sc} - i_{load} \\ L_{fc}\dfrac{dI_{fc}}{dt} = V_{fc} - \left(1-d_{fc}\right)V_{DC} \\ L_{sc}\dfrac{di_{sc}}{dt} + L_{s}\dfrac{di'_{sc}}{dt} = V_{sc} - \left(1-d_{sc}\right)V_{DC} \end{cases} \tag{5.49}$$

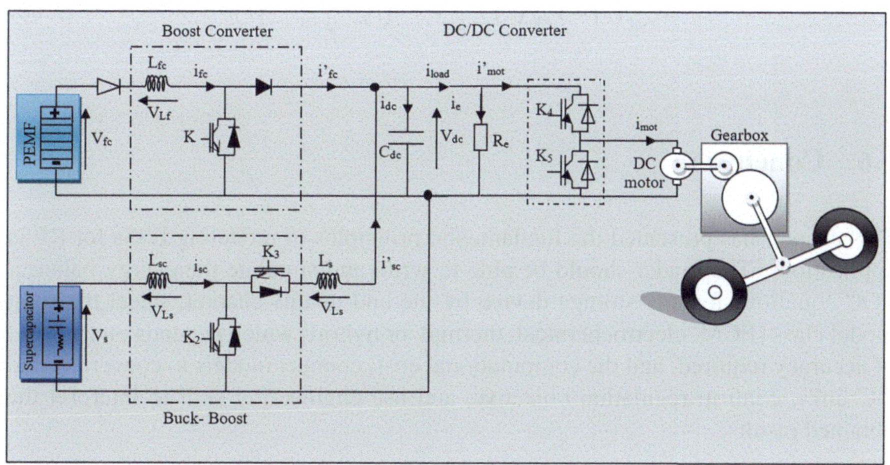

Fig. 5.17 Proposed SC/PEMFC suppling an EV

The DC motor is coupled to the wheels through a gearbox. Its mathematical model is given by the following equations:

$$\begin{cases} V_{mot}(t) = e(t) + R \, . \, i_{mot}(t) + \dfrac{L.di_{mot}(t)}{dt} \\ e(t) = K_e . \omega_{mot}(t) \\ J \dfrac{d\omega_{mot}(t)}{dt} = \mathrm{T_{em}}(\mathrm{t}) - \mathrm{T_r}(\mathrm{t}) \\ \mathrm{T_{em}}(\mathrm{t}) = K_i . i_{mot}(t) \end{cases} \tag{5.50}$$

5.5 Multi Storage Mathematical Formulation

The presented model for energy storage systems is general and flexible, capable of representing dual, hybrid, and multi-storage configurations simply by adjusting the number of storage technologies considered. By indexing each technology from 1 to *N*, the model uniformly describes the SOC dynamics, charging and discharging processes, and power balance for any combination or number of storage units.

Mathematically, for each storage technology $i \in \{1,\ldots,N\}$, the state of charge evolves over discrete time step as:

$$SOC_i(t+1) = SOC_i(t) + \frac{\eta_i^{char} \, . \, P_i^{char} \, . \, \Delta t}{E_i} - \frac{P_i^{disc}(t) \, . \, \Delta t}{\eta_i^{disc} . E_i} \tag{5.51}$$

The overall power balance at each time step is maintained by:

$$P_{load}(t) = P_{PV}(t) + \sum_{i=1}^{N} P_i^{disc}(t) - \int_{i=1}^{N} P_i^{char}(t) \tag{5.52}$$

5.6 Conclusion

This chapter has presented the fundamental principles of modeling ESSs for RESs applications. The reader should be able to write and simulate the energy balance/ SOC equations for any storage device by the end of this chapter, select the right model class (ECM, electrochemical, thermal, or hybrid) which depends on the level of accuracy required, and the computational cost; connect models to converters and DC links, confirm regulation objective, and use engineering skill to interpret the obtained results.

References

1. Xue R, Zhang G, Li Z (2025) Home energy systems in Europe: advancements and future directions. Chin J Elect Eng 11(2):38–62. https://doi.org/10.23919/CJEE.2025.000137
2. Rekioua D, Matagne E (2012) Optimization of photovoltaic power systems: modeling, simulation and control. Springer. https://doi.org/10.1007/978-1-4471-2403-05
3. Rekioua D, Roumila Z, Rekioua T (2008) Étude d'une centrale hybride photovoltaïque–éolien–diesel. J Renew Energ *11*(4):623–633. https://doi.org/10.54966/jreen.v11i4.112
4. Rekioua D, Bensmail S, Bettar N (2014) Development of hybridphotovoltaic-fuel cell system for stand-alone application. Int J Hyd Energy 39(3):1604–1611. https://doi.org/10.1016/j.ijhydene.2013.03.040
5. Balaji B, Hemalatha P, Rampradesh T, Anbarasi G, Eswari A (2024) Modelling and approximating renewable energy systems using computational intelligence. In: Computational intelligence: theory and applications. Wiley, pp 175–188. https://doi.org/10.1002/9781394214259.ch8
6. Rekioua D (2024) Wind power electric systems: modeling, simulation and control. Springer. https://doi.org/10.1007/978-3-031-52883-5
7. Ramos HM, Coronado-Hernández OE, Besharat M, Carravetta A, Fecarotta O, Pérez-Sánchez M (2025) Energy storage systems in micro-grid of hybrid renewable energy solutions. Technologies 13(11):527. https://doi.org/10.3390/technologies13110527
8. Kampker A, Späth B, Song X, Wang D (2025) Modelling of battery energy storage systems under real-world applications and conditions. Batteries 11(11):392. https://doi.org/10.3390/batteries11110392
9. Rekioua D (2023) Energy storage systems for photovoltaic and wind systems: a review. Energies 16(9):3893. https://doi.org/10.3390/en16093893
10. Abbey C, Strunz K, Joos G (2009) A knowledge-based approach for control of two-level energy storage for wind energy systems. IEEE Trans Energy Conver 24(2):539–547. https://doi.org/10.1109/TEC.2008.2001453
11. Fenner GP, Stringini LW, Rangel CAS, Canha LN (2021) Comprehensive model for real battery simulation responsive to variable load. Energies 14(11):3209. https://doi.org/10.3390/en14113209
12. Aissou R, Rekioua T, Rekioua D, Tounzi A (2016) Application of nonlinear predictive control for charging the battery using wind energy with permanent magnet synchronous generator. Int J Hydrogen Energy 41(45):20964–20973. https://doi.org/10.1016/j.ijhydene.2016.05.249
13. Chen M, Rincon-Mora GA (2006) Accurate electrical battery model capable of predicting runtime and I-V performance. IEEE Trans Energy Convers 21(2):504–511. https://doi.org/10.1109/TEC.2006.874229
14. Netchinda S, Duangpummet P (2025) Development of a measurement instrument to assess students' conceptual understanding in three levels of chemical representation of galvanic cells by using the Rasch model. In: 2025 10th International STEM Education Conference (iSTEM-Ed). IEEE, pp 1–6. https://doi.org/10.1109/iSTEM-Ed65612.2025.11129316
15. Madani SS, Shabeer Y, Allard F, Fowler M, Ziebert C, Wang Z, Panchal S, Chaoui H, Mekhilef S, Dou SX, See K, Khalilpour K (2025) A comprehensive review on lithium-ion battery lifetime prediction and aging mechanism analysis. Batteries 11(4):127. https://doi.org/10.3390/batteries11040127
16. Nath A, Mehta R, Gupta R, Bahga SS, Gupta A, Bhasin S (2022) Control-oriented physics-based modeling and observer design for state-of-charge estimation of lithium-ion cells for high current applications. IEEE Trans Control Syst Technol 30(6):2466–2479. https://doi.org/10.1109/TCST.2022.3152446
17. Turner J (2016) AutoLion™: a thermally coupled simulation tool for automotive Li-Ion batteries (2013-01-1522). In: Progress in modeling and simulation of batteries. SAE, pp 13–18
18. Pesaran A (2016) Thermal management modeling for avoidance of thermal runaway conditions in lithium ion batteries (2014-01-0707). In: Lithium ion batteries in electric drive vehicles. SAE, pp 75–82

19. Zhang Z, Zhang R, Liu X, Zhang C, Sun G, Zhou Y, Yang Z, Liu X, Chen S, Dong X, Jiang P, Sun Z (2024) Advanced state-of-health estimation for lithium-ion batteries using multi-feature fusion and KAN-LSTM hybrid model. Batteries 10(12):433. https://doi.org/10.3390/batteries10120433
20. Mannapperuma V, Gaddala LC, Zheng R, Kim D, Kim Y, Ullal A, Zhu S, Ha KP (2025) Electro-thermal modeling and parameter identification of an EV battery pack using drive cycle data. Batteries 11(9):319. https://doi.org/10.3390/batteries11090319
21. Rekioua D, Mezzai N, Mokrani Z et al (2024) Effective optimal control of a wind turbine system with hybrid energy storage and hybrid MPPT approach. SciRep 14:30013. https://doi.org/10.1038/s41598-024-78847-9
22. Kakouche K, Oubelaid A, Mezani S, Rekioua D, Rekioua T (2023) Different control techniques of permanent magnet synchronous motor with fuzzy logic for electric vehicles: analysis, modelling, and comparison. Energies 16(7):3116. https://doi.org/10.3390/en16073116
23. Turner J (2016) AutoLion™: a thermally coupled simulation tool for automotive Li-ion batteries. In: Progress in modeling and simulation of batteries. SAE, pp 13–18
24. Pesaran A (2016) Thermal management modeling for avoidance of thermal runaway conditions in lithium ion batteries. In: Lithium ion batteries in electric drive vehicles. SAE, pp 75–82
25. Lysenko O, Kuznietsov M, Hutsol T, Mudryk K, Herbut P, Vieira FMC, Mykhailova L, Sorokin D, Shevtsova A (2023) Modeling a hybrid power system with intermediate energy storage. Energies 16(3):1461. https://doi.org/10.3390/en16031461
26. Trinh H-A, Phan V-D, Truong H-V-A, Ahn KK (2022) Energy management strategy for PEM fuel cell hybrid power system considering DC bus voltage regulation. Electronics 11(17):2722. https://doi.org/10.3390/electronics11172722
27. Valizadeh M, Shiri M, Sarvenoee AK et al (2024) A comprehensive scheme for power management of FC/SC/battery, and solar-roof PV source in electric vehicle systems. Scientif Rep 14:27621. https://doi.org/10.1038/s41598-024-79241-1
28. Bensmail S, Rekioua D, Azzi H (2015) Study of hybrid photovoltaic/fuel cell system for stand-alone applications. Int J Hyd Energy 40(39):13820–13826. https://doi.org/10.1016/j.ijhydene.2015.04.013
29. Rekioua D, Rekioua T, Soufi Y (2015) Control of a grid connected photovoltaic system. In: 2015 International Conference on Renewable Energy Research and Applications (ICRERA). IEEE, pp 1382–1387. https://doi.org/10.1109/ICRERA.2015.7418634
30. Rekioua D, Kakouche K, Babqi A, Mokrani Z, Oubelaid A, Rekioua T, Azil A, Ali E, Alaboudy AHK, Abdelwahab SAM (2023) Optimized power management approach for photovoltaic systems with hybrid battery-super capacitor storage. Sustainability 15(19):14066. https://doi.org/10.3390/su151914066
31. Rekioua D, Mokrani Z, Kakouche K et al (2024) Coordinated power management strategy for reliable hybridization of multi-source systems using hybrid MPPT algorithms. Scientif Rep 14:10267. https://doi.org/10.1038/s41598-024-60116-4
32. Hassani H, Rekioua D, Zaouche F, Bacha S (2019) Supervision of hybrid renewable energy systems. In: 2019 1st International Conference on Sustainable Renewable Energy Systems and Applications (ICSRESA). IEEE, pp 1–5. https://doi.org/10.1109/ICSRESA49121.2019.9182478
33. Halmous A, Oubbati Y, Lahdeb M et al (2025) Optimizing control and management of hybrid power system, consisting PV-wind and battery-super capacitor, using COOT algorithm. Scientif Rep 15:33342. https://doi.org/10.1038/s41598-025-12585-4

Chapter 6
Energy Management Control of RESs with ESSs

6.1 Introduction

Optimizing the performance of RESs with storage is among the most important challenges in developing sustainable power solutions. Power flow coordination, generation-storage synergy, state-of-charge maintenance, and smooth integration with loads or the grid are all components of effective energy management control [1–5]. With a particular focus on hybrid renewable energy systems with storage, this chapter explores the fundamental principles, creative strategies, and advanced techniques. Renewable systems can achieve improved sustainability, greater flexibility, and superior reliability by understanding current energy management concepts and applying them to practical applications.

6.2 Control of RESs with ESSs

Optimizing the performance of RESs with ESSs requires effective energy management control. Power flow management, generating and storage unit coordination, state-of-charge level maintenance, and smooth integration with the load or grid are all part of the EMS objectives. With an emphasis on hybrid configurations, this chapter examines the concepts, approaches, and methods related to the energy management control RESs with storage [6–9]. RESs have the potential to improve sustainability, flexibility, and reliability by understanding and learning modern energy management controls for implementation.

D. Rekioua, *Energy Storage for Renewable Energy Systems*, Green Energy and Technology, https://doi.org/10.1007/978-3-032-19589-0_6

6.3 Renewable Energy Sources and Hybrid Systems

The global transition to sustainable power systems depends heavily on renewable energy sources. For example, wind power is frequently combined with hydropower and PV panels, in hybrid renewable systems, combining their complementary generation profiles. These hybrid systems offer several advantages, including improved power supply stability, increased total energy efficiency, and increased reliability by compensating for the irregular nature of individual sources [6].

6.4 The Choice of Energy Storage Technologies Used in RESs

ESSs types in RESs include a number of important technologies, each appropriate for various needs and applications. The choice is mostly based on the particular application, the quantity of storage time required, and integration with RE sources in order to efficiently manage supply and demand while improving system stability and reliability.

6.5 Energy Management Strategies of Hybrid RESs with Storage

Energy Management Systems (EMS) are essential to hybrid RESs because they allow efficient coordination between different energy sources, storage devices, and fluctuating load needs. A carefully designed control strategy maximizes the use of renewable energy, increases system reliability, and extends battery life [6].

The main techniques and approaches used in hybrid RESs for its energy management are (Fig. 6.1):

- SOC-Based Control (state of charge monitoring): which ensures battery operation within safe limits.
- Power Sharing Algorithms: to decide when each source/storage is active,
- MPPT Integration: for optimal renewable energies (PV/Wind) utilization,
- Load Prioritization: to ensures critical loads are always supplied.
- Hybrid control techniques: as FLC, MPC, etc.

6.5.1 Objectives of Energy Management in Hybrid RESs

The primary objectives are to maximize the use of renewable energy and provide an uninterrupted supply of power by optimizing the use of all RE sources and ESSs, reducing operating costs and emissions, and prolonging the lifetime of equipment.

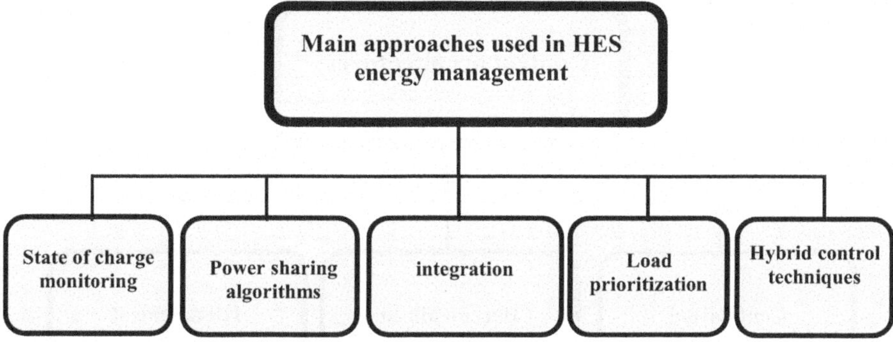

Fig. 6.1 Main approaches used in hybrid RESs energy management

Energy management aims to balance supply and demand, reduce deep discharge or component overcharging, and integrate grid connections as needed. It also helps to smooth load profiles and enable efficient energy flow between different sources and ESS.

6.5.2 Control Architectures

Performance, reliability, and scalability of microgrids and hybrid renewable energy systems are significantly influenced by the control architecture. Centralized, decentralized, and hierarchical control are the three primary control strategies that are frequently used (Fig. 6.2). Considering optimization, communication needs, and fault tolerance, each method has unique benefits and drawbacks.

- Centralized control: all components are managed by a single central controller, which maximizes system performance but requires extensive connectivity and presents a single point of failure risk.
- Decentralized control: Several local controllers operate independently, at the cost of losing global optimization or with limited communication to improve fault tolerance, scalability, and reliability.
- Hierarchical control: integrates both approaches with higher-level supervisory control coordinating overall system objectives while lower-level controllers manage local operations for responsiveness and flexibility.

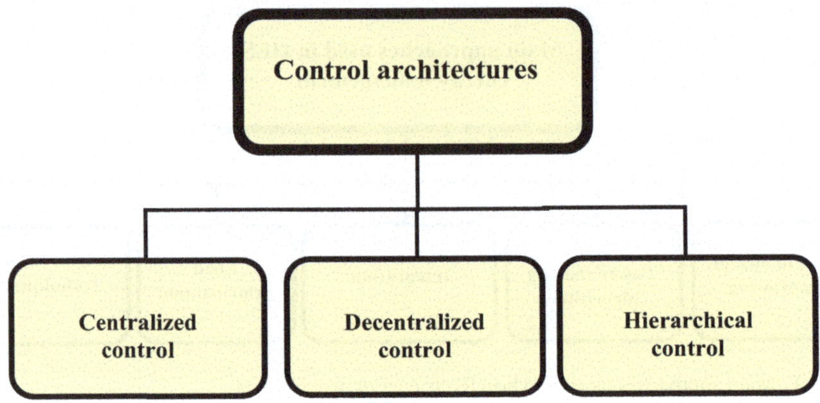

Fig. 6.2 Control architectures for hybrid RESs

6.5.3 Main Approaches of Energy Management Used in RESs

6.5.3.1 Power Flow Management and Load Balancing

Energy management systems control the real-time power flow between renewable generators, storage devices, and loads. They carefully charge or discharge storage while balancing generation output and demand to prevent excess or deficits. By ensuring peak shaving, demand changing, and smoothing fluctuation from intermittent sources, load balancing techniques decrease system component stress and improve power quality and reliability.

6.5.3.2 SOC Management and Forecasting

In order to prevent overcharging or deep discharge, which can damage batteries or other storage devices and reduce their lifespan, *SOC* management monitors the energy stored. Forecasting techniques predict future renewable generation and load demand based on weather and usage patterns, enabling proactive energy scheduling and optimized dispatch of resources. To optimize efficiency and maintain system stability, forecasting and accurate *SOC* management are essential [6–12].

This control's advantages include extending battery life, maintaining system reliability during variations and fluctuations, and ensuring availability for emergency load supply. Overall, we have:

$$SOC_{\min} \leq SOC \leq SOC_{\max} \tag{6.1}$$

Where: SOC_{min} usually set around 20–30% to prevent deep discharge, and SOC_{max}: set around 90–95% to avoid overcharging.

Different control strategies can be proposed based on this logic algorithm:

- If SOC < SOC_{min} stop discharging and prioritize charging.
- If SOC > SOC_{max} stop charging and use surplus energy elsewhere.

6.5.3.3 Power Sharing Algorithms

The objective is to distribute power generation and storage discharge across different sources in an optimal way. Various power sharing strategies can be used. It can be a Priority-Based power flow, a Proportional load sharing, droop control, … (Fig. 6.3).

- **Priority-Based Power Flow**: Power sources and storage units are ranked in priority-based power flow according to factors including cost, availability, and efficiency. Lower priority units supply the remaining loads in a planned approach once the highest priority source satisfied the power demand. Although this approach is easy to adopt, it might not optimally use all available resources [13].

Example

This is an application of Priority-Based Power Flow in a hybrid RESs with a load, diesel generator, battery, and PV system. The power supply to the load is managed in a predetermined priority order: the PV system supplies the load power first, the battery discharges to supply the remaining load if PV generation is insufficient, and the diesel generator is used as a backup source if both PV and battery cannot fully meet the demand.

This can be described as:

- Step 1: directly supply the load using PV power. The battery is charged by any excess PV energy.
- Step 2: discharge the battery to compensate the difference when PV power is less than the load demand.

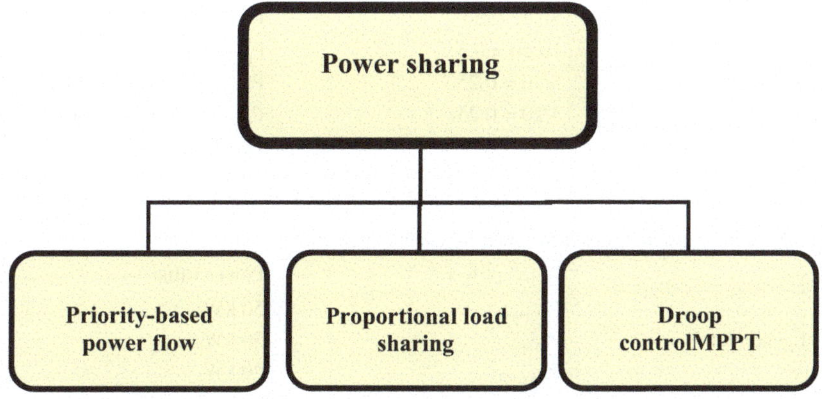

Fig. 6.3 Power sharing algorithms

- Step 3: turn on the diesel generator to satisfy the remaining demand if the PV and battery alone cannot supply the load.

- **Proportional load sharing:** This approach balances the load power among the available power sources according to defined share ratios or their capacities. This approach provides [14]:
  - Smooth system operation even under variable renewable conditions,
  - Improved reliability and lifetime of energy units,
  - Balanced use of components, preventing overloading of one source.

Example
A 10 kW photovoltaic generator, a 5 kW wind turbine system, and a 5 kW battery storage system make up an autonomous PV/Wind/batteries system. The total generation capacity is 20 kW. The power-sharing controller provides the 12 kW load demand proportionately to the sources' rated capacities (Table 6.1):

The controller dynamically redistributes with the deficit among the other sources and maintains the proportional ratios based on current availability if one source (such as wind) operates below its nominal capacity because of low wind speed.

- **Droop control**: Often used in microgrids, droop control allows decentralized power sharing without communication by simulating the natural frequency and voltage droop characteristics of synchronous generators [15].

Example
Consider an islanded AC bus (Table 6.2), connected to three distributed generators (DG):

The nominal frequency, f_0 is 50.00 Hz. Every unit is given a droop so that the frequency would decrease by Δf_{full} = 0.5 Hz at full load (a typical design option for small microgrids).

The droop slopes (in kW/Hz) under that assumption are:

Table 6.1 Share ratio and provided power of each source

Sources	Share ratio	Provided power
PV	10/20 = 0.5	$P_{pv} = 0.5 \times 12 = 6$ kW
Wind turbine	5/20 = 0.25	$P_{wind} = 0.25 \times 12 = 3$ kW
Battery	5/20 = 0.25	$P_{batt} = 0.25 \times 12 = 3$ kW
Total	**1**	**12 kW**

Table 6.2 Share ratio and provided power of each source

Sources	Rated power	Power value
PV	$P_{PV,rated}$	50 kW_P
Diesel generator	$P_{DG,rated}$	30 kW_P
Battery	$P_{Bat,rated}$	20 kW_P

$$\begin{cases} K_{PV} = \dfrac{P_{PV,rated}}{\Delta f_{full}} \\ K_{DG} = \dfrac{P_{DG,rated}}{\Delta f_{full}} \\ K_{batt} = \dfrac{P_{batt,rated}}{\Delta f_{full}} \end{cases} \tag{6.2}$$

For each unit, the droop law (active power vs. frequency) is linearized as follows:

$$\begin{cases} \Delta P_{PV} = -K_{PV} \, . \, \Delta f \\ \Delta P_{DG} = -K_{DG} \, . \, \Delta f \\ \Delta P_{batt} = -K_{batt} \, . \, \Delta f \end{cases} \tag{6.3}$$

Where:

$$\Delta f = f - f_0 \tag{6.4}$$

A positive ΔP_i (increase in supplied power) results from a negative Δf (frequency drop).

Let's suppose there is a sudden increase in the load of: ΔP_{load} = +30 kW (for example, a massive motor starts off). This additional 30 kW must be instantly supplied by the grid. The frequency will decrease to a new steady value Δf that provides power balance when the generators operate independently under droop control.

$$\begin{aligned} \Delta P_{load} &= \sum_i \Delta P_i = \Delta P_{PV} + \Delta P_{DG} + \Delta P_{batt} \\ &= -K_{PV} \, . \, \Delta f - K_{DG} \, . \, \Delta f - K_{batt} \, . \, \Delta f \\ &= -. \, \Delta f \left(K_{PV} + K_{DG} + K_{batt} \right) \\ &= -\Delta f \, . \, K_{load} \end{aligned} \tag{6.5}$$

We determine the frequency deviation Δf:

$$\Delta f = -\frac{\Delta P_{load}}{K_{load}} \tag{6.6}$$

$$\Delta f = -\frac{30}{200} + 60.15 Hz$$

And:

Table 6.3 Power distribution among distributed energy sources under droop control after a load increase

Source	Rated power (kW)	(K_i) (kW/Hz)	ΔP_i contribution (kW)
PV	50	$K_{pv} = 50/0.5 = 100$	$\Delta P_{pv} = 100 \times 0.15 = 15$
Diesel	30	$K_{DG} = 30/0.5 = 60$	$\Delta P_{DG} = 60 \times 0.15 = 9$
Battery	20	$K_{batt} = 20/0.5 = 40$	$\Delta P_{batt} = 40 \times 0.15 = 6$
Total	100	$K_{total} = 100 + 60 + 40 = 200$	$\Delta P_{total} = 30$

$$\mathrm{f} = f_0 + \Delta f \tag{6.7}$$

So the steady frequency becomes

$$\mathrm{f} = 50 - 0.15 = 49.85Hz$$

Table 6.3 summarizes the different calculations for each source.

It is noticed that $\Delta P_{total} = 30$ kW matches the required step $\Delta P_{load} = +30$ kW.

Therefore, assuming their pre-step outputs were at an initial value P_{i0}, the new operating powers become:

$$\begin{cases} P_{PV,new} = P_{PV} + \Delta P_{PV} \\ P_{DG,new} = P_{DG} + \Delta P_{DG} \\ P_{batt,new} = P_{batt} + \Delta P_{batt} \end{cases} \tag{6.8}$$

Their contributions are precisely the ΔP_i numbers above if we assume they were operating close to zero excess margin before the step.

6.5.3.3.1 MPPT Strategies

Maximizing energy extraction from wind turbines or PV panels in a variety of environmental conditions is the objective. Various MPPT techniques can be applied to PV (Perturb & Observe (P&O), Incremental Conductance (INC), Fuzzy Logic or Neural Network-based MPPT, etc.) [16–19] and wind turbines (Power Signal Feedback (PSF), Tip-Speed Ratio (TSR), etc.) [6, 20–22]. If the power increases ($\Delta P < 0$), the perturbation direction is reversed and the system has moved away from the MPP. Both the wind and PV subsystems can converge toward the MPP using the iterative Perturb and Observe (P&O) method. ΔP_{PV} and ΔP_{Tb} represent the variations in PV and turbine power after each perturbation; and sign(ΔP) denotes the direction of power change (Fig. 6.4).

The MPPT adjustment is controlled by the following equation:

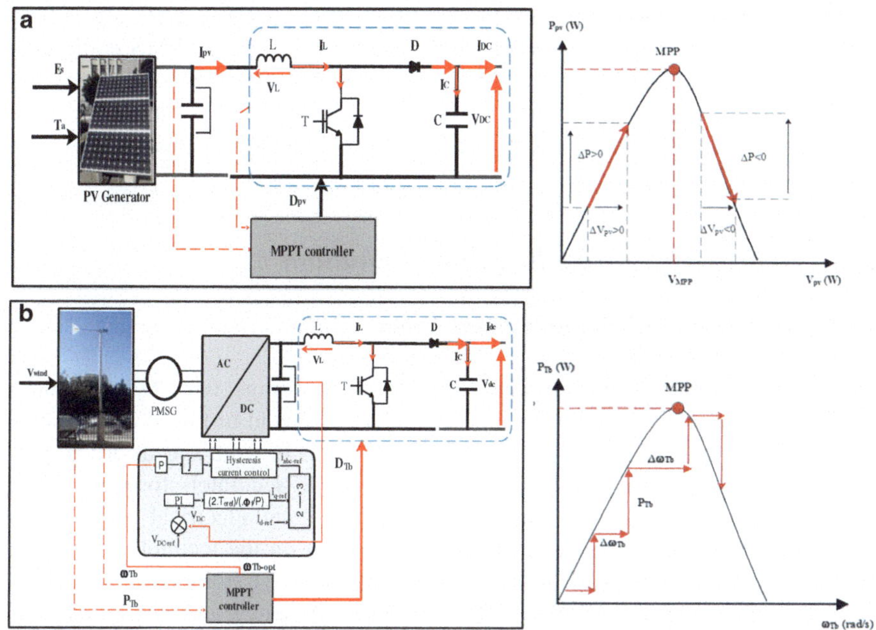

Fig. 6.4 MPPT controllers operation. (**a**) In PV system, (**b**) In wind turbine system

$$\begin{aligned} D_{PV}(\mathrm{k}+1) &= D_{PV}(\mathrm{k}) + K_{D_{PV}}\ .\ sign(\Delta P_{PV}) \\ D_{Tb}(\mathrm{k}+1) &= D_{Tb}(\mathrm{k}) + K_{D_{Tb}}\ .\ sign(\Delta P_{Tb}) \end{aligned} \tag{6.9}$$

6.5.3.3.2 Integration with Energy Management System (EMS)

Highly effective and smart energy management in renewable microgrids is made possible by the combination of MPPT and EMS. By modifying operating points in response to system needs and environmental conditions, MPPT controllers maximize the output from solar panels or wind turbines, ensuring that the EMS continuously receives the maximum amount of power possible [13, 22–24].

By continuously optimizing energy extraction from renewable sources like PV arrays or wind turbines, the MPPT system plays an essential role in supporting the EMS. For example, in PV systems, the MPPT dynamically modifies the operating voltage and current to guarantee that the system continuously provides the EMS with the maximum amount of power available, independent of changes in temperature or solar irradiation. Fundamental data, such as power, voltage, current, and operational status, can be sent in real time between the MPPT and EMS via integrated protocols for data acquisition and monitoring [25–27].

MPPT controllers with fuzzy logic or artificial intelligence (AI) algorithms improve tracking accuracy and overall performance in advanced EMS configurations. After that, the EMS uses this smart input to generate data-driven, adaptive control decisions that assure optimal power distribution, better energy use, and improved system performance under variable environmental conditions.

6.5.4 Hybrid RESs Modeling and Simulation

6.5.4.1 Modeling of RESs and Storage Units

To design and optimize the performance of RESs using storage, wind-hybrid system modeling and simulation are necessary. Characterizing wind turbines' aerodynamics, power curves, and dynamic response to fluctuations in wind speeds are all part of modeling wind turbine and PV power generation. To accurately forecast output power, accurate models incorporate turbine mechanical and electrical behaviors as well as stochastic wind speed fluctuations. Depending on the technology, models depict the electrical, chemical, or mechanical properties of storage units, such as flywheel mechanical inertia, fuel cell or electrolyser efficiency, and battery state-of-charge dynamics.

6.5.4.2 Simulation Tools

MATLAB/Simulink, HOMER, PSCAD, Open Modelica, and other simulation programs and platforms are commonly used in RESs [18, 28, 29]. These platforms offer large libraries and modular components for modeling energy storage devices (batteries, supercapacitors, hydrogen systems), advanced control algorithms (MPPT, EMS, or predictive control), and renewable energy generation systems. By combining electrical, thermal, and control subsystems into a single environment, they enable thorough system simulations. Before putting the system into use, engineers can use these simulations to evaluate dynamic behavior, load management techniques, and energy optimization scenarios [30–32].

6.5.4.3 Case Studies Demonstrating Energy Management Benefits

Various renewable and storage technologies have been shown by a number of case studies and simulation-based evaluations. For example, by balancing instantaneous power fluctuations between generation and load, the EMS is important in managing wind power variations and preserving power quality in hybrid systems that integrate wind turbines with battery storage. Fuel consumption and operating cost are reduced as a result of this integration, which also eliminates generator damage, lowers voltage and frequency fluctuations, and minimizes the need for diesel backup systems.

Furthermore, the adoption of AI-based forecasting models and predictive control strategies to predict wind variations and improve storage device management is increasing [31]. By decreasing excessive charge-discharge cycles, these intelligent control solutions have been demonstrated to considerably reduce energy losses, maximize renewable production, and extend the lifespan of storage devices.

6.6 Challenges and Future Perspectives

6.6.1 Technical Challenges in Control and Management

Because renewable energy sources are intermittent and variable, managing hybrid RESs is a complex challenge. Their output is mostly dependent on irregular environmental factors that change over minutes or even seconds, such as temperature, wind speed, solar irradiation, and weather fluctuation. This fluctuation complicates real-time control, load balancing, and precise forecasting to keep power quality. Intelligent energy management algorithms and advanced control are needed for this.

EMS use sophisticated algorithms to address these problems. Hybrid systems can operate in an adaptive, robust, and predictive way. Based on real conditions and forecasted demand, these algorithms dynamically distribute energy among renewable sources, storage devices (batteries, supercapacitors, hydrogen, etc.), and backup generators. While maintaining grid regulation, they aim to reduce power fluctuations, maximize the effective use of storage devices, and increase the utilization of RESs.

Practical hardware limitations, like as inverter response, converter efficiency limits, and battery deterioration, make it more difficult to achieve these objectives. Real-time synchronization between subsystems may be affected by communication delays in distributed control networks.

A multidisciplinary approach is necessary for the effective management of hybrid renewable systems. This involves integrating secure communication infrastructures that ensure resilience against disturbances, robust hardware and control architectures for real-time responsiveness, and intelligent algorithms for decision-making [30–32].

6.6.2 Integration with Smart Grids and Demand Response

Advanced communication protocols, compatibility standards, and coordinated control mechanisms with grid operators are necessary for the smooth integration of hybrid RESs into smart grids in order to obtain reliable and effective operation. Communication protocols enable real-time data exchange between storage systems, distributed energy resources, and the grid. Interconnection enables effective communication between a range of technologies, including PV inverters, wind converters, battery management systems, and electric vehicles, using specified data formats and control interfaces [33, 34].

6.6.3 Emerging Technologies and Research Directions

In order to improve forecasting, fault detection, and adaptive control, future research is exploring artificial intelligence and machine learning [32]. Emerging energy storage technologies that promise greater longevity and efficiency include supercapacitors, and improved hydrogen storage. In order to maximize resource utilization, sector coupling, which integrates transportation, heating, and energy is increasingly incorporated into hybrid system design. Demand response programs, in which the Energy Management System (EMS) adjusts power generation or consumption in response to grid signals, allow hybrid systems to actively participate. This coordination enables load shifting, peak shaving and frequency regulation.

6.7 Applications

6.7.1 Application of Single Storage in PV, Wind and Hybrid RESs

6.7.1.1 PV Systems with ESS

For instance, Fig. 6.5 illustrates the interaction between a PV array, MPPT controller, battery storage, and EMS.

The system's brain is the EMS. It gets information from the MPPT and monitors the batteries' SOC. The EMS determines when to store excess energy in batteries through a buck-boost converter or when to deliver power from the batteries to the load during low generation times.

Optimal energy extraction, steady DC bus voltage (V_{DC}) management, and intelligent power distribution among PV generating, storage, and the load are all ensured by this configuration [33–37]. By combining fast MPPT dynamics with the EMS's strategic decision-making capabilities, the system improves the PV installation's

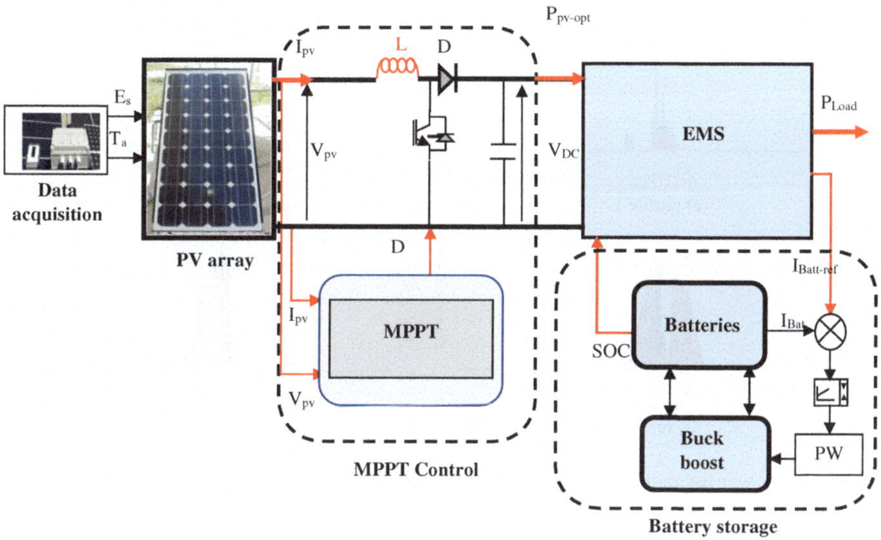

Fig. 6.5 Example of MPPT with EMS for PV/battery system

Table 6.4 Different cases in PV with battery storage

PV/Batteries	**P_{load}**	P_{pv}	0	$P_{pv} + P_{Batt}$	P_{pv}	P_{pv}	0	P_{Batt}	0
	DP	=0	<0	<0	>0	>0	<0	<0	<0

reliability and effectiveness. The various operating modes (M1–M8) of the PV–Battery–EMS hybrid system shown in Table 6.4.

$$\Delta P = \mathrm{P}_{load} - P_{PV} \tag{6.10}$$

The PV power obtained for the measured solar irradiance for four different days. The load is a water pumping system which is an important application [35–37], where the pump power has been designed first before the simulation study. The proposed management algorithm is given in Fig. 6.6.

The proposed management algorithm for PV system with batteries supplying water pump is shown in Fig. 6.7. Also the different powers are represented in Fig. 6.8. This figure highlights how the battery storage smooths out fluctuations in PV output, ensuring more consistent power availability despite daily irradiance variability. The different power flows in the PV-battery system show how excess PV power charges the battery and how the battery discharges (in red color) to meet load demands when solar input is insufficient. This demonstrates the battery's role in balancing supply and demand, enhancing system reliability.

The SOC variations is shown in Fig. 6.9. The SOC profiles over four days confirm that the battery operates within safe limits without deep discharges or overcharging. This effective power management prolongs battery life and ensures

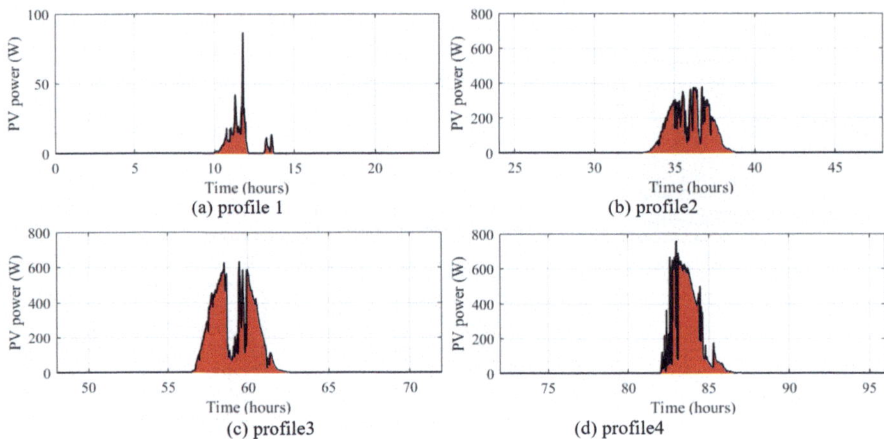

Fig. 6.6 PV power for each different profile. (**a**) Profile1, (**b**) Profile2, (**c**) Profile3, (**d**) Profile4

availability of stored energy during low solar periods, critical for continuous pump operation.

6.7.1.2 Wind Turbine/Battery System

Figure 6.10 depicts the configuration of a wind turbine with battery storage and an EMS or power management control (PMC) for optimum power flow. Despite the irregular and discontinuous nature of wind energy, this structure aims to provide steady operating, effective energy use, and a constant power supply [25–27].

The Electrical power is significantly influenced by the turbine's features and wind speed profile that is currently available. To ensure efficient operation, the system employs the MPPT algorithm, which continuously adjusts the generator's operating point to maximize power from the available wind. The MPPT controller ensures that the turbine operates on its optimal power curve even when the wind is changing quickly by regulating the duty cycle (D_{cycle}) of the converter.

The PMC or EMS chooses the best operating mode from a number of potential scenarios based on this information:

- The PMC uses a bidirectional DC–DC converter to charge the battery when wind power exceeds the load needs.
- In order to ensure that the load has a steady supply, the PMC activates the battery to discharge (boost mode) when wind generation is insufficient.
- The PMC may limit power extraction or redirect excess energy in order to protect system components in situations with strong wind or full battery capacity.

Power stability and energy storage are the two functions of the battery storage system. By delivering power during low wind or intermittent conditions and absorbing energy surpluses during strong wind times, it significantly decreases output

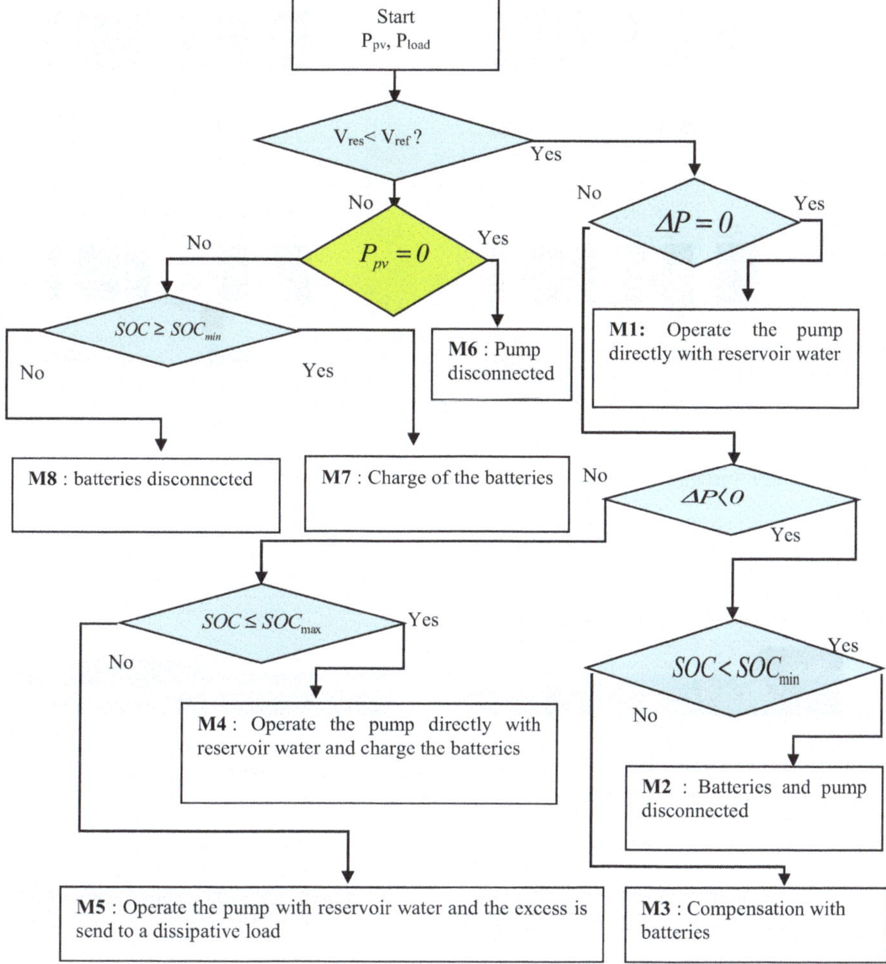

Fig. 6.7 Proposed management algorithm for PV system with batteries supplying water pump

variations. This improves power quality and reduces the strain on the mechanical and electrical subsystems of the turbine.

Through this coordinated control of the MPPT and PMC, the system offers intelligent power distribution between the wind generator, storage, and the load, optimal energy extraction, and DC bus voltage stability. The system's autonomy and dependability are also increased by incorporating energy storage, which enables it to function well in isolated or remote locations.

The proposed management algorithm for wind turbine system with batteries supplying water pump is shown in Fig. 6.11. And the various operating modes (M1–M8) of the wind–Battery–EMS hybrid system shown in Table 6.5.

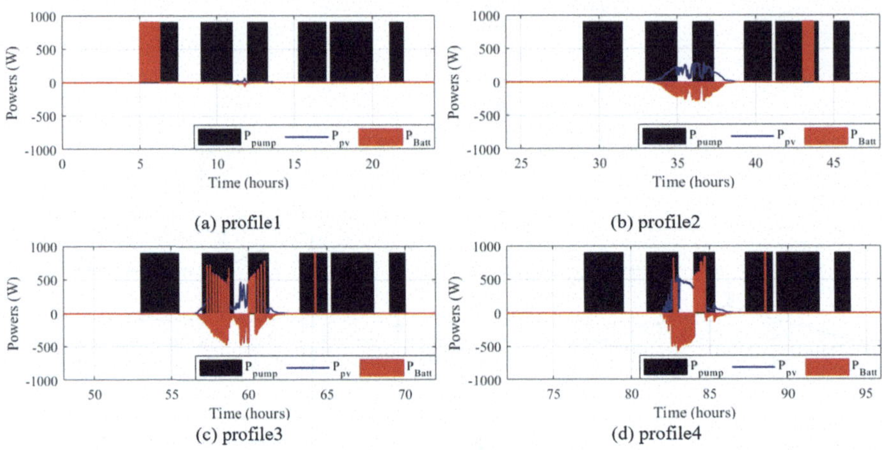

Fig. 6.8 Power variations for each different profile. (**a**) Profile1, (**b**) Profile2, (**c**) Profile3, (**d**) Profile4

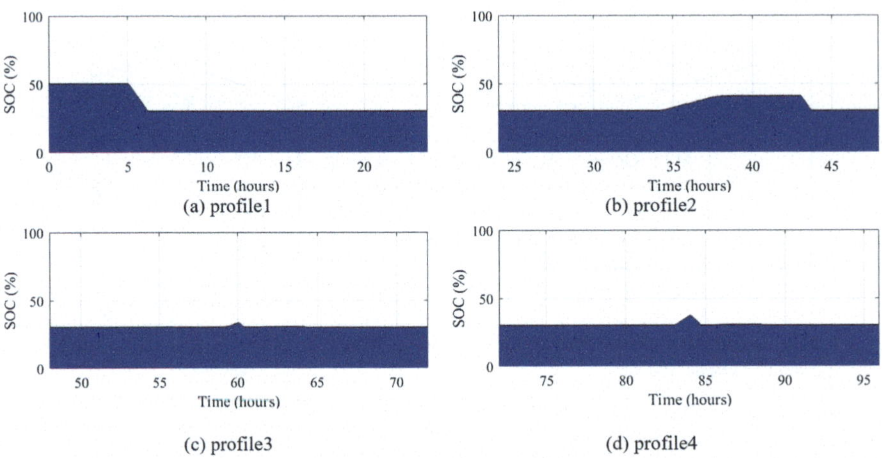

Fig. 6.9 SOC variations for each different profile. (**a**) Profile1, (**b**) Profile2, (**c**) Profile3, (**d**) Profile4

Wind power output fluctuates according to measured wind speeds, which vary significantly day-to-day (Fig. 6.12). Battery storage again plays a fundamental role in mitigating the intermittency of wind energy, providing a regulator that stabilizes

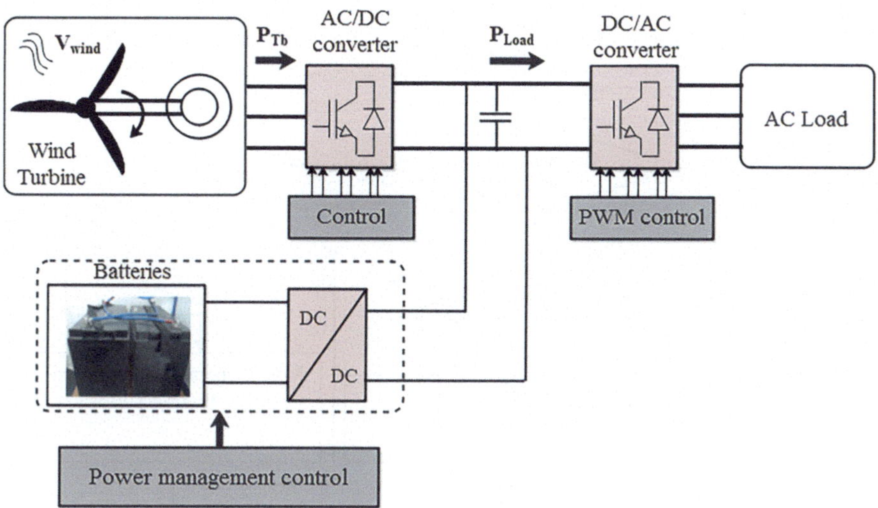

Fig. 6.10 Wind turbine with battery storage and PMC

power delivery to the pump. Figure 6.13 also depicts the various powers in the case 2. The battery's relationship to wind power generation and pump demand is depicted in this image (in black).

Figure 6.14 displays the SOC variations. The SOC is observed to remain within ideal operating ranges, indicating that the battery is successfully managed to prevent stress. This is crucial for the system's durability and dependability, particularly in situations with variable wind.

$$\Delta P = P_{load} - P_{wind} \tag{6.11}$$

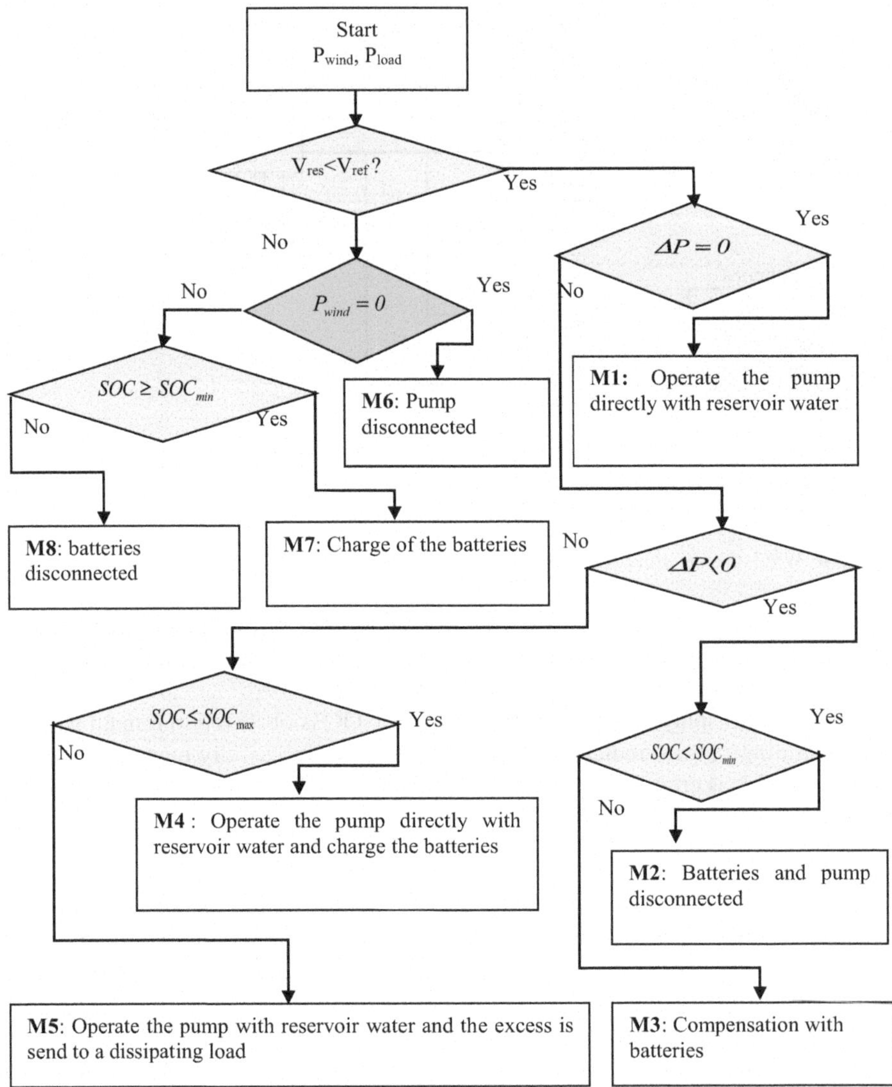

Fig. 6.11 Proposed management algorithm for wind turbine system with batteries supplying water pump

Table 6.5 Different modes obtained in wind turbine system with battery and PMC

Wind Turbine/batteries	P_{Load}	P_{wind}	0	$P_{wind} + P_{Batt}$	P_{wind}	P_{wind}	0	P_{Batt}	0
	DP	=0	<0	<0	>0	>0	<0	<0	<0

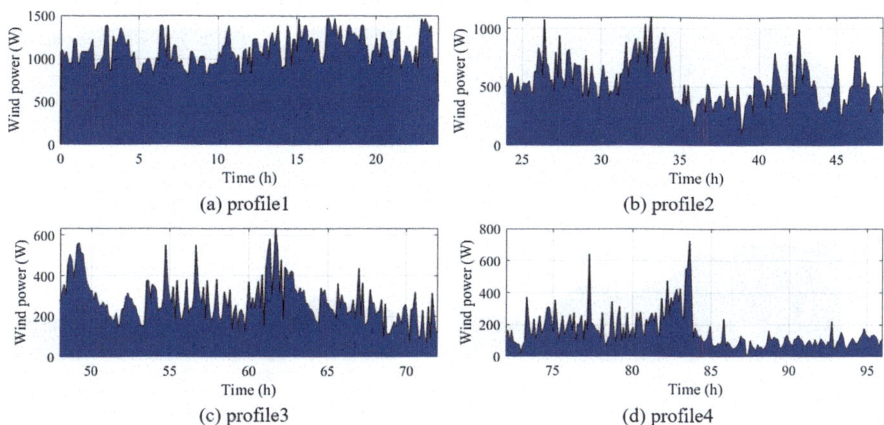

Fig. 6.12 Wind turbine power during different profiles. (**a**) Profile1, (**b**) Profile2, (**c**) Profile3, (**d**) Profile4

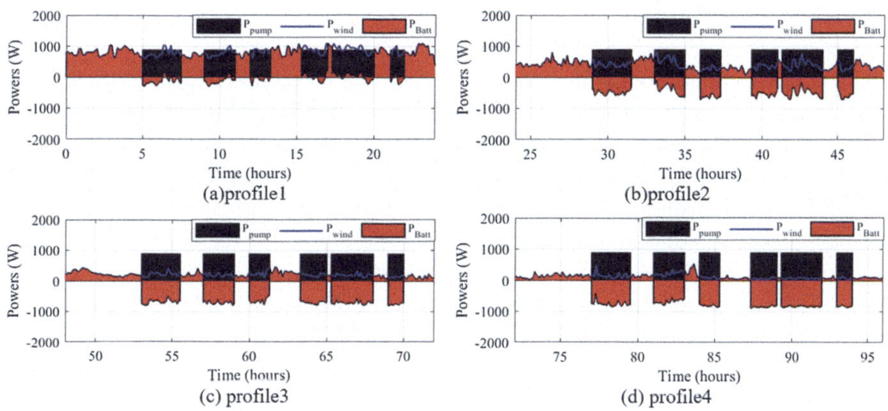

Fig. 6.13 Power variations during different profiles. (**a**) Profile1, (**b**) Profile2, (**c**) Profile3, (**d**) Profile4

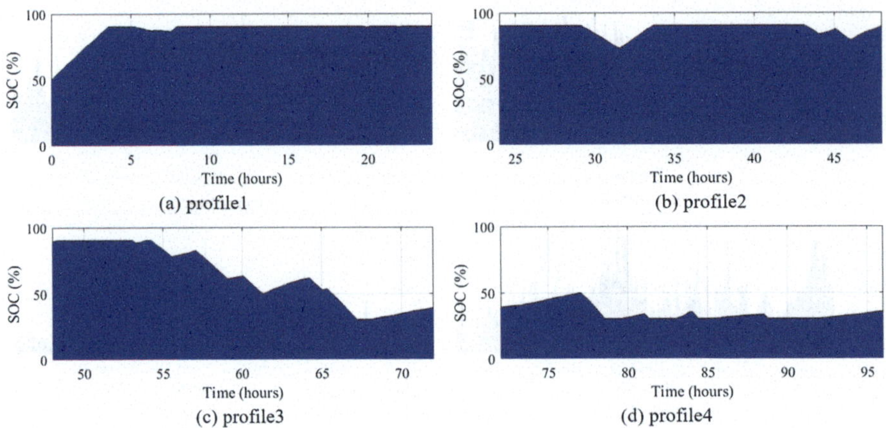

Fig. 6.14 SOC variations during different profiles. (**a**) Profile1, (**b**) Profile2, (**c**) Profile3, (**d**) Profile4

6.8 Conclusion

The solution to maximizing the potential of hybrid RESs with ESSs is EMS control. This chapter covered a variety of approaches, including advanced integration of MPPT with EMS, power sharing, and state-of-charge monitoring. These methods provide an uninterrupted supply of energy, increase system efficiency, and encourage sustainable development by optimizing energy flows and smartly coordinating various generation and storage units. Even if there are still technical obstacles to overcome, such as intermittency, hardware constraints, and communication delays, new intelligent algorithms and robust architectures promise adaptive, predictive management. Even if there are still technical challenges like intermittency, hardware limitations, and communication delays, new intelligent algorithms and robust architectures promise adaptive, predictive control that can get over these obstacles.

References

1. Gevorkov L, Domínguez-García JL, Romero LT (2022) Review on solar photovoltaic-powered pumping systems. Energies 16(1):94. https://doi.org/10.3390/en16010094
2. Rekioua D (2024) Wind power electric systems: modeling, simulation and control. Springer. https://doi.org/10.1007/978-3-031-52883
3. Rekioua D, Matagne E (2012) Optimization of photovoltaic power systems: modeling, simulation and control. Springer. https://doi.org/10.1007/978-1-4471-2403-05
4. Guerrero JM, Vasquez JC, Matas J, de Vicuña LG, Castilla M (2011) Hierarchical control architecture for islanded microgrids. IEEE Trans Ind Electron
5. Rekioua D (2023) Energy storage systems for photovoltaic and wind systems: a review. Energies 16(9):3893. https://doi.org/10.3390/en16093893
6. Rekioua D, Kakouche K, Babqi A, Mokrani Z, Oubelaid A, Rekioua T, Azil A, Ali E, Alaboudy AHK, Abdelwahab SAM (2023) Optimized power management approach for photovoltaic

systems with hybrid battery-supercapacitor storage. Sustainability 15(19):14066. https://doi.org/10.3390/su151914066
7. Pragya, Thakur R (2022) A review of architecture and control strategies of hybrid AC/DC microgrid. In: 2022 International Conference on Intelligent Controller and Computing for Smart Power (ICICCSP). IEEE, pp 1–5. https://doi.org/10.1109/ICICCSP53532.2022.9862386
8. Qian Z, Cao S, Jiang H, Krishnan SB, Tan KT, Tseng KJ (2022) Optimized EMS real-time simulation and validation for Singapore Pulau Ubin microgrid. In: 2022 12th International Conference on Power and Energy Systems (ICPES). IEEE, pp 401–406. https://doi.org/10.1109/ICPES56491.2022.10073143
9. Zheng Y, Jia J, An D (2025) Energy management for microgrids with hybrid hydrogen-battery storage: a reinforcement learning framework integrated multi-objective dynamic regulation. Processes 13(8):2558
10. Shete S, Jog P, Kumawat RK, Palwalia DK (2021) Battery management system for SOC estimation of lithium-ion battery in electric vehicle: a review. In: 2021 6th IEEE International Conference on Recent Advances and Innovations in Engineering (ICRAIE). IEEE, pp 1–4. https://doi.org/10.1109/ICRAIE52900.2021.9703752
11. Xiong R, Shen W (2019) Battery state of charge and state of energy estimation. In: Advanced battery management technologies for electric vehicles. Wiley, pp 67–93. https://doi.org/10.1002/9781119481652.ch3
12. Zhao R, Tang L (2025) Research on energy-saving adaptive optimization of hybrid electric vehicle based on improved dynamic programming and control rule extraction. Sci Rep 15:41564. https://doi.org/10.1038/s41598-025-25431-4
13. Rekioua D, Roumila Z, Rekioua T (2008) Étude d'une centrale hybride photovoltaïque–éolien–diesel. J Renew Energies 11(4):623–633. https://doi.org/10.54966/jreen.v11i4.112
14. Arise N, Bhoomika V, Reddy NA, Harika S, Koushik A (2023) Power generation of wind-PV-battery-based hybrid energy system for standalone AC microgrid applications. In: 2023 5th International Conference on Smart Systems and Inventive Technology (ICSSIT). IEEE, pp 261–266. https://doi.org/10.1109/ICSSIT55814.2023.10060963
15. Peng H, Wang D, Su M (2018) Comparative study of power droop control and current droop control in DC microgrid. In: The 11th IET international conference on Advances in Power System Control, Operation and Management (APSCOM 2018). IET, pp 1–4. https://doi.org/10.1049/cp.2018.1783
16. Buraimoh E, Aluko AO, Oni OE, Davidson IE (2022) Decentralized virtual impedance-conventional droop control for power sharing for inverter-based distributed energy resources of a microgrid. Energies 15(12):4439. https://doi.org/10.3390/en15124439
17. Rekioua D (2020) MPPT methods in hybrid renewable energy systems. In: Hybrid renewable energy systems. Springer, pp 79–138. https://doi.org/10.1007/978-3-030-34021-6_3
18. Rekioua D, Mezzai N, Mokrani Z et al (2024) Effective optimal control of a wind turbine system with hybrid energy storage and hybrid MPPT approach. Sci Rep 14:30013. https://doi.org/10.1038/s41598-024-78847-9
19. Aissou R, Rekioua T, Rekioua D, Tounzi A (2016) Application of nonlinear predictive control for charging the battery using wind energy with permanent magnet synchronous generator. Int J Hydrog Energy 41(45):20964–20973. https://doi.org/10.1016/j.ijhydene.2016.05.249
20. Belaid S, Rekioua D, Oubelaid A, Ziane D, Rekioua T (2022) Proposed hybrid power optimization for wind turbine/battery system. Periodica Polytech Electr Eng Comput Sci 66(1):60–71. https://doi.org/10.3311/PPee.18758
21. Rekioua D, Mokrani Z, Kakouche K et al (2024) Coordinated power management strategy for reliable hybridization of multi-source systems using hybrid MPPT algorithms. Sci Rep 14:10267. https://doi.org/10.1038/s41598-024-60116-4
22. Cao F, Jiang W, Ma R, Zhou Y, Bai H, Chen J (2025) Optimal sizing algorithm of multi source hybrid power system for unmanned aerial vehicle. In: IECON 2025 – 51st annual conference of the IEEE industrial electronics society. IEEE, pp 1–6. https://doi.org/10.1109/IECON58223.2025.11221961

23. Magableh MAK, Radwan AAA, Mohamed YA-RI, El-Saadany EF (2024) A novel reduced-order modeling approach of a grid-tied hybrid photovoltaic–wind turbine–battery energy storage system for dynamic stability analysis. IEEE Open J Power Electron 5:1459–1483. https://doi.org/10.1109/OJPEL.2024.3455933
24. Saravanan K, Ramaswamy A (2022) Feasibility analysis for sustainable energy generation from medium and large-scale industries. In: 2022 international conference on computer, power and communications (ICCPC). IEEE, pp 468–473. https://doi.org/10.1109/ICCPC55978.2022.10072116
25. Tadjine K, Rekioua D, Belaid S, Rekioua T, Logerais PO (2022) Design, modeling and optimization of hybrid photovoltaic/wind turbine system with battery storage: application to water pumping. Math Model Eng Probl 9(3):655–667. https://doi.org/10.18280/mmep.090312
26. Datta R, Ranganathan V (2002) DFIG wind turbine control. IEEE Trans Power Syst
27. Belaid S, Rekioua D, Oubelaid A, Ziane D, Rekioua T (2022) A power management control and optimization of a wind turbine with battery storage system. J Energy Storage 45:103613. https://doi.org/10.1016/j.est.2021.103613
28. Rekioua D, Zaouche F, Hassani H, Rekioua T, Bacha S (n.d.) Modeling and fuzzy logic control of a stand-alone photovoltaic system with battery storage. Turk J Electromech Energy 4(1):11–17. https://sloi.org/urn:sl:tjoee41137
29. Abderezzak B, Rekioua D, Binns R, Busawon K, Hinaje M, Douine B et al (n.d.) Technical feasibility assessment of a PEM fuel cell refrigerator system. Int J Hydrog Energy 45(19):11211–11219. https://doi.org/10.1016/j.ijhydene.2018.04.060
30. Khotsriwong N, Boonraksa P, Sarapan W, Boonraksa T, Boonrakchat N, Marungsri B (2023) Short-term load demand forecasting using supervised deep learning techniques: a case study of Suranaree university of technology. In: 2023 international electrical engineering congress (IEECON). IEEE, pp 417–420. https://doi.org/10.1109/iEECON56657.2023.10126570
31. Aldulaimi MYM, Çevik M (2025) AI-enhanced MPPT control for grid-connected photovoltaic systems using ANFIS-PSO optimization. Electronics 14(13):2649. https://doi.org/10.3390/electronics14132649
32. Joo K, Woo SS (2021) IVDR: imitation learning with variational inference and distributional reinforcement learning to find optimal driving strategy. In: 2021 20th IEEE international conference on machine learning and applications (ICMLA). IEEE, pp 256–262. https://doi.org/10.1109/ICMLA52953.2021.00047
33. Dargan J, Gupta N, Singh L (2021) Blockchain based energy management system: a proposed model. In: 2021 international conference on technological advancements and innovations (ICTAI). IEEE, pp 510–514. https://doi.org/10.1109/ICTAI53825.2021.9673233
34. Rekioua D, Bensmail S, Bettar N (2014) Development of hybrid photovoltaic-fuel cell system for stand-alone application. Int J Hydrog Energy 39(3):1604–1611. https://doi.org/10.1016/j.ijhydene.2013.03.040
35. Rahrah K, Rekioua D, Rekioua T, Bacha S (2015) Photovoltaic pumping system in Bejaia climate with battery storage. Int J Hydrog Energy 40(39):13665–13675. https://doi.org/10.1016/j.ijhydene.2015.04.048
36. Ould-Amrouche S, Rekioua D, Hamidat A (2010) Modelling photovoltaic water pumping systems and evaluation of their CO_2 emissions mitigation potential. Appl Energy 87(11):3451–3459. https://doi.org/10.1016/j.apenergy.2010.05.021
37. Adjati A, Rekioua T, Rekioua D (2021) Use of the dual stator induction machine in photovoltaic-wind hybrid pumping. J Européen des Systèmes Automatisés 54(1):115–124. https://doi.org/10.18280/jesa.540113

Chapter 7
Optimization, Integration, and Strategic Perspectives of RESs

7.1 Introduction

For renewable energy sources to be successfully used in power grids, ESSs are essential drivers. Advanced optimization approaches are necessary to design, control, and operate an ESS under complex and often conflicting system limitations in order to achieve their full potential. The nonlinear, multi-objective nature of these problems is addressed in this chapter by exploring optimization techniques based on mathematics, analysis, heuristics, and artificial intelligence. These techniques, which range from dynamic programming to classical linear programming and further, offer a solid basis for balancing longevity, system reliability, cost, and efficiency [1–5].

7.2 Optimization Approaches for ESSs

7.2.1 Mathematical and Analytical Optimization Approaches

In renewable energy applications, the design of ESSs largely depends on mathematical and analytical optimization techniques. Under specific technological and environmental constraints, these methods aim to identify the best configuration, operation, and control strategies that either minimize costs or maximize system performance. They offer an extensive structure for simulating the complicated interactions that exist between loads, storage devices, and RE resources. Several optimization strategies are employed to increase the efficiency of RESs with integrated storage [6–13] (Fig. 7.1).

Table 7.1 provides a description of different approaches.

D. Rekioua, *Energy Storage for Renewable Energy Systems*, Green Energy and Technology, https://doi.org/10.1007/978-3-032-19589-0_7

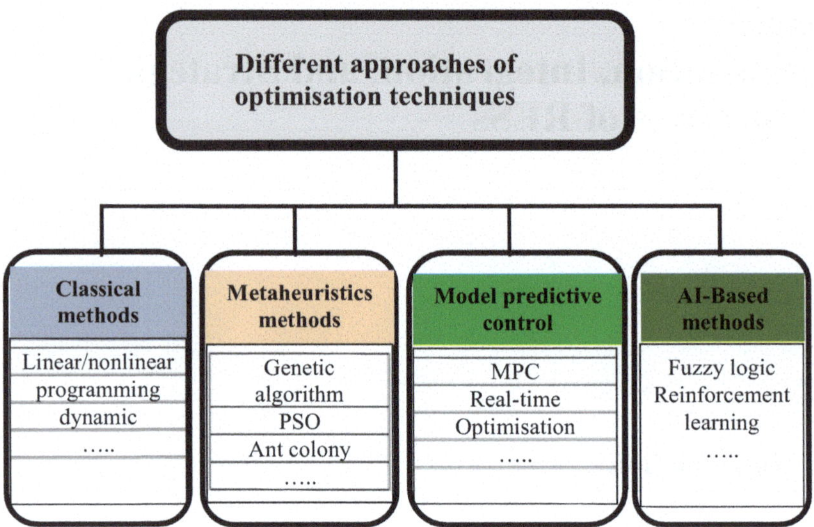

Fig. 7.1 Different approaches of optimisation techniques

Table 7.1 Description of optimization techniques

Optimization Method	Description
Classical methods	Provide mathematically rigorous solutions that provide optimal control and cost effectiveness for systems.
Metaheuristic algorithms	Provide flexibility and efficiency in resolving complex, nonlinear, and multi-objective optimization situations.
Model predictive control	Predicts future system performance and dynamically modifies control actions to maintain optimal performance under changing operating conditions.
AI-based methods	Incorporate intelligence and flexibility into control strategies to enable systems to learn from data, manage uncertainty.

7.2.1.1 Classical Methods

7.2.1.1.1 Linear and Nonlinear Optimization Models

ESS sizing and operation frequently employs both linear and nonlinear optimization techniques. Energy storage system (ESS) sizing and operation planning are clearly addressed in Fig. 7.2, which highlights the combination of linear and nonlinear optimization techniques.

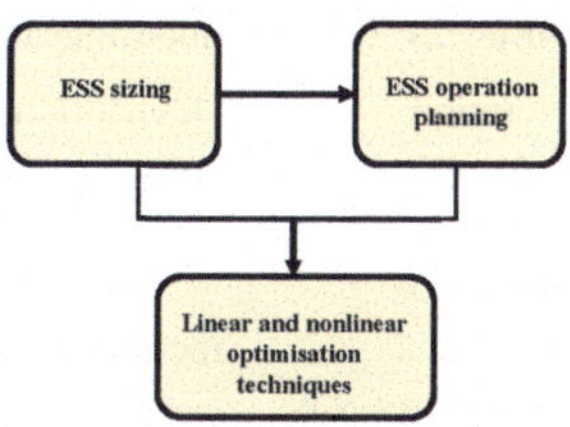

Fig. 7.2 ESS sizing and operation planning using linear and nonlinear optimization techniques

Table 7.2 Examples of nonlinearities in renewable energy conversion and storage systems

Source	Cause of nonlinearity
Converter efficiency	Efficiency is affected by operating voltage and load.
Battery degradation	Internal resistance vary with temperature, aging, and charge/discharge rate.
Fuel cell performance	Ohmic losses and activation have an impact on the voltage-current relationship.
Wind turbine characteristics	The aerodynamic power coefficient $C_p(\lambda,\beta)$.

7.2.1.1.2 Nonlinear Programming (NLP)

Numerous system components in practical renewable energy systems have nonlinear characteristics that make linear equation modeling unfeasible. Energy conversion losses, fluctuating converter and storage device efficiency, and dynamic interactions between subsystems (as wind turbine/battery/fuel cell coupling) are some of the causes of these nonlinearities. Nonlinear Programming (NLP) techniques are therefore required to capture the actual behavior of the system and to achieve more realistic optimization results. Table 7.2 summarizes the typical nonlinearities in energy systems.

A general nonlinear optimization formulation can be written as:

Minimise f(x) subject to:

$$\begin{cases} g_i(x) \leq 0 & i = 1,2,\ldots.m \\ h_j(x) \leq 0 & j = 1,2,\ldots.n \end{cases} \tag{7.1}$$

The main notations used to formulate the hybrid energy system optimization issue are shown in Table 7.3.

Application

For the battery charge/discharge scheduling optimization in a wind/FCs/battery system that optimizes operating costs while taking battery deterioration and converter efficiency into account, this is an example of a mathematical formulation. The formulation specifies the notation and decision variables (Table 7.4), the objective function, the constraints, and possible solution approaches, then to find an objective function f(x) to minimize total cost, write all necessary constraints (limits, battery dynamics, balancing) and finally suggestions for solutions.

Table 7.3 General notations and descriptions in optimization problem formulation

<table>
<tr><th>Notation</th><th colspan="3">Description</th></tr>
<tr><td>f(x)
What we want to minimize</td><td colspan="3">Objective function</td></tr>
<tr><td>$g_i(x)$
What must stay within limits</td><td colspan="3">Inequality constraints</td></tr>
<tr><td>$h_j(x)$
What must always hold true</td><td colspan="3">Equality constraints</td></tr>
<tr><td rowspan="5">x
What we can control</td><td rowspan="5">Vector of decision variables, such as converter duty cycles, energy flows, or state of charge (SOC).</td><td>P_{gen}</td><td>Generated power</td></tr>
<tr><td>P_{stor}</td><td>Charging and discharging storage power</td></tr>
<tr><td>SOC</td><td>Battery state of charge</td></tr>
<tr><td>V, I</td><td>Voltage and current variables</td></tr>
<tr><td>D_{cycle}</td><td>Converter duty cycle</td></tr>
</table>

Table 7.4 Notation and decision variables in hybrid wind/FCs/battery system

Parameter	Description	Parameter	Description
$SOC(t)$	Battery state of charge	η_{conv}	Efficiency of the converter
$P_{wind}(t)$	Wind generation (given /forecast)	$E_{batt\text{-}cap}$	Nominal battery energy capacity
$P_{load}(t)$	Electrical demand	SOC_{min}, SOC_{max}	Minimum and maximum state of charge
$C_{FC}(t)$	Marginal cost of producing 1 kWh with the fuel cell, includes fuel and O&M	$P_{FC\text{-}max}$, $P_{FC\text{-}min}$	Maximum and minimum fuel cell power output
$c_{grid}(t)$	Grid electricity price (if grid exchange allowed)	$P_{batt\text{-}ch,max}$, $P_{batt\text{-}dis,max}$	Maximum battery charge and discharge power limits
$\eta_{charge}, \eta_{discharge}$	Battery charge/discharge efficiencies	c_{deg}	Battery degradation cost coefficient
$P_{FC}(t)$	Fuel cell electrical output	$P_{grid\text{-}imp}(t)$, $P_{grid\text{-}exp}(t)$	Grid import and export power
$P_{batt\text{-}ch}(t)$	Battery charge power (≥0).	$P_{batt\text{-}dis}(t)$	Battery discharge power

7.2.1.1.3 Linear Programming (LP)

In order to provide dependable and cost-effective system operation, this optimization achieves a balance between energy generation, storage use, and cost effectiveness. The mathematical formulation is to minimize the total cost over the time horizon T:

$$C_{total} = \sum_{t=1}^{T} \left(C_{gen} . P_{gen,t} + C_{stor} . P_{stor,t} \right) \quad (7.4)$$

The power balance is:

$$P_{load,t} = P_{gen,t} + P_{stor,t} \quad (7.5)$$

The storage device's maximum power capacity is specified by the storage operating limits.

$$0 \leq P_{stor,t} \leq P_{stor,\max} \quad (7.6)$$

Application

In order to minimize overall operating expenses while preserving power balance, the application uses a linear programming (LP) technique to optimize the wind/battery system's charging and discharging schedule over a 24-h period. Due to system limitations, the total power from the battery $P_{stor,t}$ and the wind generator $P_{gen,t}$ must meet the load requirement, with the battery's charging and discharging rate being restricted to a maximum of 4 kW. In particular, the battery must discharge $P_{stor,t} = 2$ kW to meet the demand if the load at hour *t* is 5 kW and the wind generator supplies 3 kW.

Table 7.5 provides examples of the schedules.

The LP algorithm carefully determines these charging and discharging choices in order to save overall costs, avoid exceeding the battery charge and discharge limitations, and guarantee that the system remains balanced at each time step.

Table 7.6 shows the operational choices that the LP algorithm would optimize for cost reduction while maintaining that load requirements are always satisfied.

This can be represented in Fig. 7.3:

7.2.1.1.4 Dynamic Programming for Storage Scheduling

When system states change over time in successive processing of difficulties, dynamic programming (DP) is a useful optimization method. When determining the best charge/discharge intervals for energy storage systems with fluctuating renewable generation and demand profiles, it is particularly useful. The principle of DP is

Table 7.5 Provided schedule of the example

Hours	Wind power	Load power	Battery power
1–5	2 kW	3 kW	Discharges 1 kW
6–10	5 kW	3 kW	Charges 2 kW
11–15	4 kW	4 kW	No charge/discharge
16–24	2 kW	3 kW	Discharges 1 kW

Table 7.6 Operating schedule of wind/battery system with power balance and storage constraints

Time (h)	Wind power	Load power	Battery action
1–5	**2** kW	$P_{load} = P_{wind} + P_{batt} = 2 + 1 = \mathbf{3}$ kW	Battery discharges **1** kW
6 – 10	**5** kW	$P_{load} = P_{wind}\text{-}P_{batt\text{-}charge} = 5 - 2 = \mathbf{3}$ kW	Battery charges **2** kW
11–15	**4** kW	$P_{load} = P_{wind} = 4$ kW	Balanced battery idle
16–24	**2** kW	$P_{load} = P_{wind} + P_{batt\text{-}dis} = 2 + 1 = \mathbf{3}$ kW	Battery discharges **1** kW

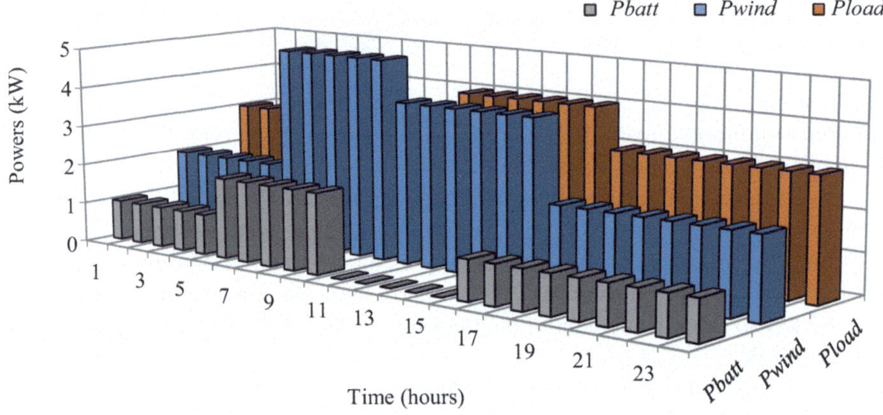

Fig. 7.3 Obtained powers in wind/battery system

to decomposes a complicated optimization problem into a number of smaller, more manageable subproblems [14–16]. Every stage denotes a time step (for example, hourly), and the decision variables match control operations like charging, discharging, or keeping the storage unit idle. For example, power generation and demand fluctuate over time in renewable systems (wind, solar, or hybrid).

The difficulty is determining whether the storage unit (battery or supercapacitor) should:

- Charge (store excess renewable energy),
- Discharge (provide power when renewable generation is low), or
- Remain idle (to avoid unnecessary cycling) at each time step.

Example

This is a case study of how battery storage scheduling in a PV system uses dynamic programming (DP). The objective is to determine the best charge/discharge strategy with variable PV generation and load. Let's examine the following data during a 10-h period (Table 7.7).

This, can be represented in Fig. 7.4:

The PV/battery system's dynamic behavior during a 10-h period, is determined by the MATLAB dynamic-programming simulation (Fig. 7.5).

Table 7.7 PV generation, load demand, and optimal storage action

Hour (t)	P_{pv} (kW)	P_{load} (kW)	$DP = P_{pv} - P_{load}$ (kW)	Optimal action (from DP)	*SOC* (%)
1	1	3	–2	Discharge	↓
2	2	3	–1	Discharge	↓
3	4	3	+1	Charge	↑
4	6	3	+3	Charge	↑
5	4	2	+2	Charge	↑
6	3	2	+1	Charge	↑
7	1	3	–2	Discharge	↓
8	0	4	–4	Discharge	↓
9	0	4	–4	Discharge	↓
10	0	3	–3	Discharge	↓

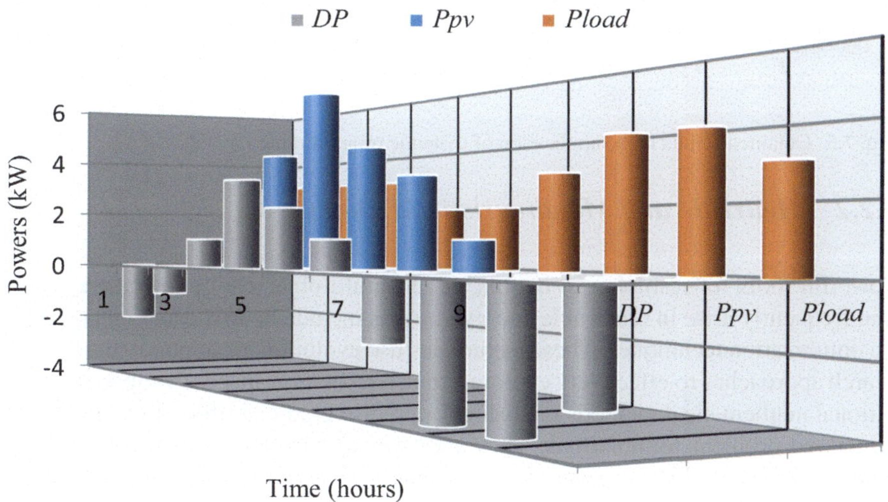

Fig. 7.4 Power flows in the PV/battery system

With a maximum PV production of about 6 kW around the fourth hour, which corresponds to peak irradiation, the PV generation and load demand profiles fluctuate naturally over time. There are alternate periods of energy excess and deficit as the load stays almost constant between 2 kW and 4 kW. When PV generation exceeded the load between hours three and six, the controller chooses the charging mode, which raises the SOC. In order to prevent needless cycling, idle operation occurs during the intermediate times when PV and load are nearly balanced. The system starts discharge mode, supplying the load with energy and reducing the SOC, if the load power exceeds the PV power (hours 1–2 and 7–10).

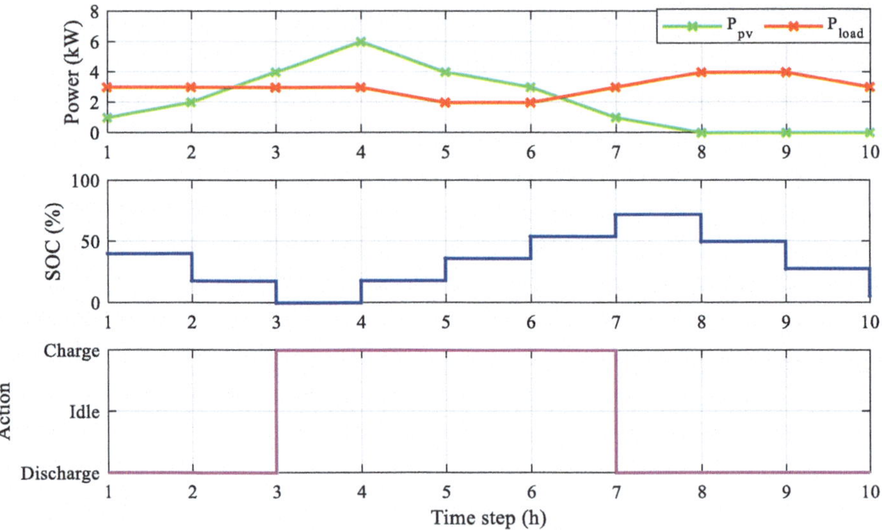

Fig. 7.5 Obtained results in the application of dynamic programming

7.2.2 *Heuristic and Metaheuristic Methods*

Powerful tools for resolving in nonlinear, and multi-objective optimization issues that frequently arise in renewable energy systems include heuristic and metaheuristic optimization techniques. These techniques use evolutionary or population-based search approaches to effectively explore extensive solution spaces, compared to traditional mathematical methods requiring for precise gradient information and well-structured problem formulations. They are especially useful for managing energy storage, designing hybrid systems, and managing energy under uncertainty when analytical models are difficult to get or computationally expensive.

Natural evolution and genetics serve as the inspiration for genetic algorithms, which use processes such as hybridization, variation, and selection, known as crossover, mutation, and selection to develop potential solutions across many generations.

The main steps are [17–20]:

- Initialization: create a starting population of potential solutions (chromosomes) that correspond to the system control variables (such as converter parameters, SOC limits, and battery power).
- Evaluation: determine the fitness function, such as total losses, cost, or system efficiency.
- Selection: based on fitness, pick the top-performing options.
- Mutation: to produce diversity, combine and slightly modify solutions.

Genetic algorithms (GA) are a useful tool for minimizing the total cost function in hybrid systems with storage, which includes the costs of generation and storage operation.

Example
This is an example of how to use genetic algorithms (GA) to understand battery management in PV systems. It shows how an intelligent controller makes hourly decisions on whether to charge, discharge, or remain the battery idle in order to balance PV generation and load demand.

Table 7.8 shows how the system is supposed to operate.

Figure 7.6 depicts daily operation of a PV/battery system, and each subplot verifies that the control method is acting exactly as expected. The battery stores extra midday solar energy and provides power during non-solar hours, confirming that GA-based control correctly maintains the energy balance.

7.2.3 *Artificial Intelligence (AI)*

Artificial intelligence (AI) and machine learning (ML) are rapidly revolutionizing the optimization of RESs by supporting predictive and adaptive decision-making. While classic optimization or heuristic approaches that rely on predetermined models or rules, AI-based solutions learn system behavior and dynamically adjust to shifting operational conditions.

Real-time control adaptation to system dynamics and uncertainties is made possible by AI-driven optimization in RESs which combine different technologies. To effectively regulate the distribution of power across storage units, different methods are often used.

The basic concepts are [21–25]:

- Adaptive decision-making**:** using real-time measurements (such as SOC, power variations, and load demand), the controller continually adjusts its parameters.
- Hybrid energy coordination: by balancing long-term energy supply (batteries or fuel cells) with rapid response (supercapacitors), the AI system decides how much power should be distributed to each storage device.
- Degradation-aware optimization*:* machine learning models can adapt control strategies to prolong lifespan by learning how storage components degrade.

Table 7.8 Expected operation

Period	P_{PV}/P_{Load}	Battery action	SOC
0–6 h	$P_{PV} < P_{Load}$	Discharge	Decreasing
7–12 h	$P_{PV} > P_{Load}$	Charge	Increasing
13 h	$P_{PV} \approx P_{Load}$	Balanced	--------------
14–24 h	$P_{PV} < P_{Load}$	Discharge	Decreasing again

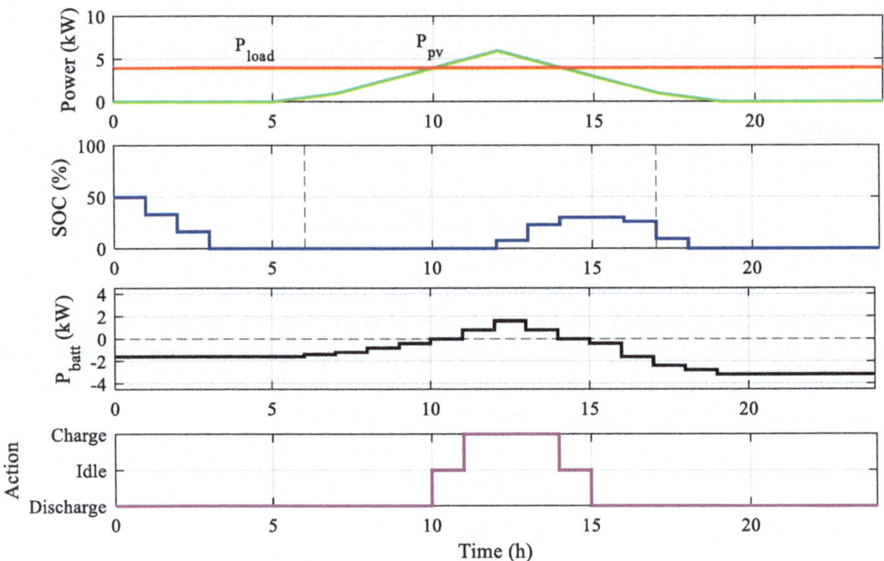

Fig. 7.6 Variations of powers and state of charge

Table 7.9 Inputs and outputs used in a fuzzy controller

Variable	Type (Input/ Output)	Description
SOC	Input	Battery state of charge
$\Delta P = P_{pv} - P_{load}$	Input	Power difference between PV generation and load
I_{batt}	Output	Battery current

Example

This is an example of FLC applied in PV/battery systems. The goal is to manage battery charge/discharge intelligently based on system state to maximize battery lifespan, maintain grid stability, and ensure energy availability.

A fuzzy controller is used and the different inputs and outputs used (Table 7.9) and fuzzy sets with membership functions (Table 7.10).

Each linguistic term is represented by a triangular or trapezoidal membership function. The most common value is represented by a triangular membership function, which has a simple peak shape with gradual transitions on both sides. Because of its simplicity and ease of computation, it is frequently utilized. A trapezoidal membership function includes sloped sides that suggest gradual transitions to near terms, and a flat top that indicates a range of values that fully belong to the term.

So, in this case, we will obtain seven rules (Table 7.11).

Table 7.10 Fuzzy sets and membership functions

Input/Output	Membership functions		
Input 1: SOC (%)	Low (**L**)	0–30	Triangular/trapezoidal
	Medium (**M**)	20–80	Triangular
	High (**H**)	70–100	Triangular/trapezoidal
Input 2: ΔP (kW)	Negative (**N**)	$P_{pv} < P_{load}$ (deficit)	
	Zero (**Z**)	$P_{pv} = P_{load}$(balanced)	
	Positive (**P**)	$P_{pv} > P_{load}$ (surplus)	
Output: Battery current (A)	Strong charge (**SC**)		
	Mild charge (**MC**)		
	Idle (**I**)		
	Mild discharge (**MD**)		
	Strong discharge (**SD**)		

Table 7.11 The different obtained rules

Rule	If (SOC)	And (ΔP)	Then (Battery Action)
R1	L	P	SC
R2	L	Z	MC
R3	M	P	MC
R4	M	Z	I
R5	M	N	MD
R6	H	P	I
R7	H	N	SD

Application 2

In this application, it is examined a hybrid renewable system that combines fuel cells (FCs), battery storage, and PV panels. The hybrid system is managed intelligently and adaptably through the use of a FLC [26–28].

The various variables which determine how the controller operates are given in Table 7.12.

Five rules can be obtained (Table 7.13).

Fuzzy inference uses the Mamdani approach, whereas defuzzification uses the centroid (center of gravity) method to convert fuzzy results into logical control actions (Fig. 7.7).

It is considered two (02) tests. The first one under a sinusoidal variation of solar irradiance variations and the second under a step profile of solar irradiance (Fig. 7.8).

It can be observed that the battery SOC and hydrogen levels can change smoothly under the sinusoidal profile because the relays *(K_1–K_5)* show fewer abrupt transitions and longer time durations in each mode. With the step profile, on the other hand, the controller quickly switches between modes, creating distinct apparent switching events in K_1–K_5 that correspond to each irradiance step. This action helps verify that the logic correctly manages changing conditions and validates the FLC rule base.

Table 7.12 Input variables and output signals

Input variables		Output signals $K_i \in \{ON, OFF\}$ ON = 1, OFF = 0	
E_s	Solar irradiance	K_1	PV relay
SOC	State of charge	K_2	Fuel cell relay
		K_3	Battery relay
TH_2	Hydrogen level	K_4	Electrolyzer relay
		K_5	Load relay or auxiliary storage control

Table 7.13 Obtained five rules

Rule	Es	SOC	TH_2	K_1	K_2	K_3	K_4	K5
R_1	L	L	L	0	0	0	0	0
R_2	M	L	M	1	1	1	0	0
R_3	H	M	L	1	0	1	0	1
R_4	H	H	H	1	0	0	1	0
R_5	M	H	M	1	0	1	0	1

7.2.4 Multi-Objective Optimization (MOO)

Increasing component lifetime, reducing overall system costs, and improving energy efficiency are just a few of some conflicting objectives that must be balanced in order to optimize energy storage-based renewable energy systems. Achieving this balance is necessary to achieve both economic and technological sustainability. The MMO aims to determine a set of optimal trade-offs (Pareto front) rather than a single solution. Each point on the Pareto front represents a distinct balance between different objectives [29, 30].

The typical goals consist of:

- Efficiency maximization: reduce losses in converters, inverters, and storage systems to increase overall energy conversion and storage efficiency.
- Minimize costs: by decreasing energy losses, capital investment (CAPEX), and operating and maintenance (O&M) costs.

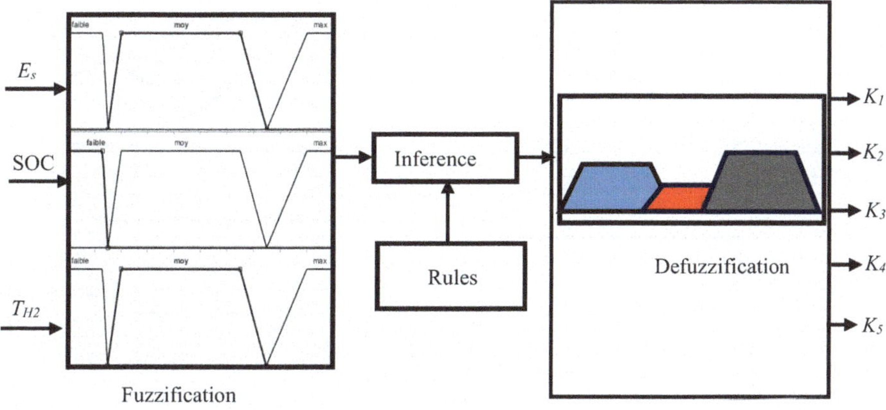

Fig. 7.7 Control system based on FLC

7.3 Economic and Environmental Assessment of Energy Storage

7.3.1 Economic Evaluation Criteria

7.3.1.1 The Levelized Cost of Storage (LCOS)

This indicator (LCOS), includes the cost of capital, operating and maintenance expenses, replacement costs, and salvage value. All of which have been adjusted to present value. It is essential for assessing the technological maturity and investment value of storage projects [31–34].

- If it is assuming full capital loss, LCOS can be expressed as:

$$LCOS = \frac{Total \quad lifetime \quad cost}{Total \quad energy \quad delivered \quad over \quad time} = \frac{C_{capex} + \sum_{t=1}^{N} \frac{C_{opex,t}}{(1+r)^t}}{E_{delivered}} \tag{7.7}$$

- It can be written by taking account the is the residual/salvage value (V_{res}):

$$LCOS = \frac{C_{capex} + \sum_{t=1}^{N} \frac{C_{opex,t}}{(1+r)^t} - V_{res}}{E_{delivered}} \tag{7.8}$$

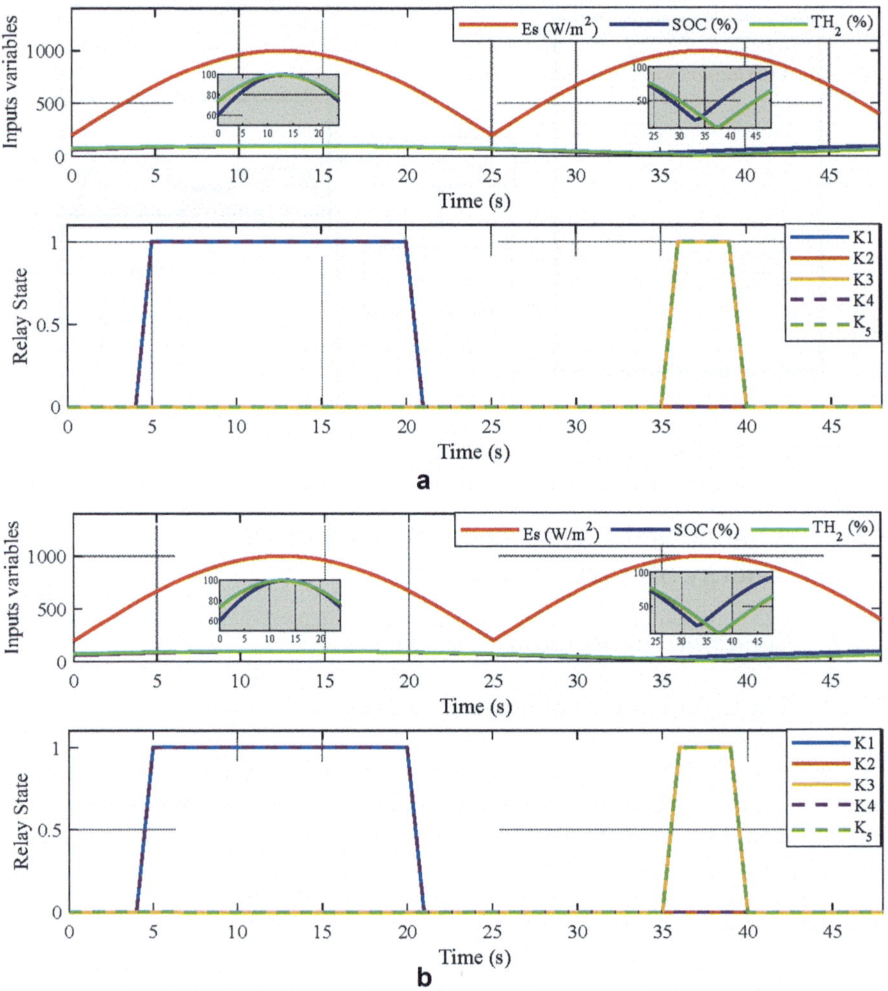

Fig. 7.8 Obtained input variables and relay states in the PV/battery/FCs example. (**a**) Under a sinusoidal variation of solar radiation variations (**b**) Under a step profile of solar radiation

Example

Let's suppose that capital costs C_{CAPEX} = 10,000 €, the discounted O&M cost is = 5,000 €, the total energy provided is 50,000 kWh, the value of residual V_{res} = 2,000€.

We will obtain:

$$LCOS = \frac{10.000 + 5.000 - 2.000}{50.000} = 0.26€ / \text{kWh}$$

Without the value left over:

$$LCOS = \frac{10.000 + 5.000}{50.000} = 0.30€ / \text{kWh}$$

7.3.1.2 Net Present Value (NPV)

It calculates how profitable an energy storage investment will be. In other words, NPV calculates an investment's overall economic value after taking the time value of money into consideration.

It is provided by:

$$NPV = \sum_{t=1}^{N} \frac{R_t - C_t}{(1+r)^t} \tag{7.9}$$

Where: R_t is the revenue or savings from the ESS, C_t is the operational cost and a positive value of NPV indicates a profitable investment.

$$C_t = C_{O\&M} + C_{replacement} + C_{losses} \tag{7.10}$$

$$C_{total} = C_{inv} + \sum_{t=1}^{T} C_t \tag{7.11}$$

7.3.1.3 Return on Investment (ROI)

ROI evaluates the long-term profitability by comparing initial and initial costs against system revenues or savings. It is defined as:

$$ROI = \frac{Total \quad net \quad profit}{Total \quad investment \quad cost}.100 \tag{7.12}$$

Or:

$$\begin{aligned} ROI &= \frac{\sum_{t=1}^{T} R_t - C_{tot}}{C_{tot}}.100 \\ &= \frac{R_{tot} - C_{tot}}{C_{tot}}.100 \end{aligned} \tag{7.13}$$

where: $R_{tot} = \sum_{t=1}^{T} R_t$ is the total cumulative revenues or savings from ESS operation over its lifetime T, and $C_{tot}$is the total capital and operational costs (CAPEX + OPEX).

7.3.1.4 Payback Period (PBP)

Depending on load profiles, tariff policies, and system use, it offers a simple but instructive measure of economic viability that usually ranges from 3 to 6 years.

$$PBP = \frac{C_{inv}}{Annual \quad net \quad savings} \tag{7.14}$$

7.3.1.5 Application

This application requests an analysis of a renewable energy project's economic viability over a ten-year (10) period, such as a solar, wind, or hybrid system (Table 7.14).

This application requires calculation of NPV, ROI, and Payback Period. After coding under Matlab we can find the following results (Table 7.15):

7.3.2 *Cost-Benefit Analysis of Different Storage Technologies*

A number of technical, financial, and environmental considerations need to be carefully considered when choosing an ESS for the integration of RE. These evaluations ensure that the chosen storage technology optimizes the economic return and

Table 7.14 Different parameters used in the economic evaluation

Parameter	Symbol / Code	Value	Description
Project lifetime	T	10 years	Duration over which the project is evaluated economically.
Discount rate	r	0.08 (8%)	Reflects the time value of money, future revenues are discounted.
Initial investment	C_{inv}	€50,000	Capital expenditure (CAPEX) at project start.
Annual revenues	R_t	[8,000, 8,200, 8,500, 8,700, 8,800, 9,000, 9,200, 9,400, 9,500, 9,700] €	Expected energy sales or cost savings each year. The values increase slightly each year (due to improved performance or rising energy prices).
Annual maintenance costs	C_t	1,000 € per year	Operation & Maintenance (O&M) costs, assumed constant.
Net annual cash flow	Cash flow = $R_t - C_t$	Varies from 7,000 to 8,700 €	Net profit after deducting annual maintenance cost.

Table 7.15 Obtained indicators value

Indicator	Value
NPV	2222.48 €
ROI	58.00%
Payback period	≈ 7 years

Table 7.16 Evaluation categories and parameters for energy storage technologies

Category	Parameters
Technical performance	Storage capacity Efficiency Response time Cycle life
Economic aspects	Initial investment cost Operation and maintenance Payback period
Environmental considerations	Material sustainability Recyclability Carbon footprint
Application-specific requirements	Grid stabilization Peak shaving Energy arbitrage Renewable energy smoothing
System integration parameters	Power electronics compatibility Control and management strategies Scalability and modularity

reduces environmental effect in addition to supporting system performance and reliability (Table 7.16).

7.3.3 Environmental Impacts and Carbon Footprint

7.3.3.1 Life-Cycle Assessment (LCA) and Sustainability Indicators

It is a comprehensive method for evaluating how an ESS will affect the environment over the life cycle. This covers the extraction of raw materials extraction, transportation, operation, and recycling at the end of life [35–37] (Table 7.17).

7.4 Integration of Energy Storage with Smart Grids and IoT

Smart grids are an example of how traditional electrical grids have developed into dynamic, flexible, and intelligent energy networks. To provide real-time energy flow monitoring, and optimization, they combine distributed generation, RE sources, and advanced communication technologies. Because they offer peak shaving, load

Table 7.17 Sustainability indicators for environmental assessment of energy storage systems

Sustainability Indicator	Description
Carbon footprint	Measurement of the overall GHG emissions related to the ESS's life cycle.
Material sustainability	Encourage the use of abundant and recyclable resources by focusing on the environmental cost and availability of the raw materials used in the system.
Recyclability of components	Evaluation of end-of-life processes, including the possibility of component recycling, which decreases pollution and extraction of resources.
Energy payback time	Time needed for the ESS's operating energy savings to balance the energy used during production and its whole life cycle.
Resource depletion	Assessment of non-renewable resource consumption throughout the system's life cycle with the objective of reducing the exploitation of limited resources.

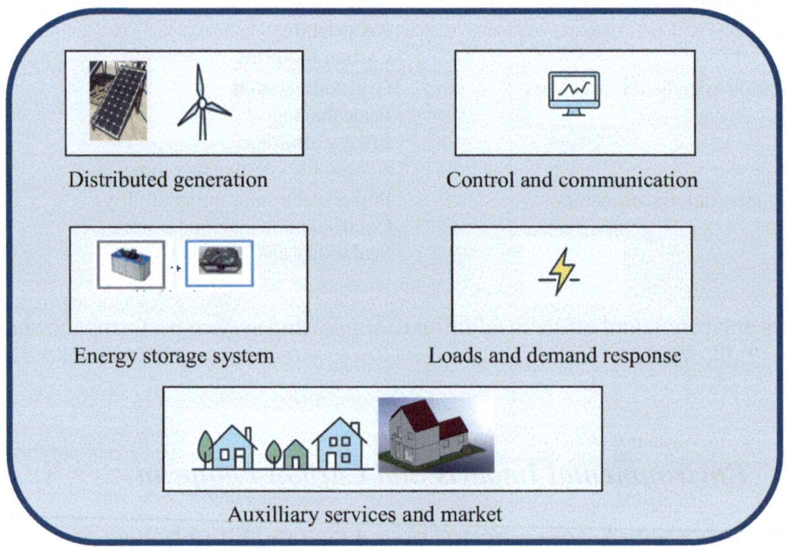

Fig. 7.9 Smart grid architecture with energy storage

shifting, frequency management, and compensation for renewable intermittency, energy storage systems (ESSs) are crucial to this transition [38–40]. ESSs promote grid stability and energy sustainability by allowing the incorporation of decentralized renewable electricity, improving grid reliability, and offering auxiliary services through coordinated operation (Fig. 7.9).

7.4.1 Concept of Smart Grids and the Role of ESSs

The goal of smart grids is to transform conventional one-way electrical networks into intelligent and smart systems that may rapidly balance supply and demand. A smart grid incorporates digital communication, automation, and distributed generation (PV, wind, micro-turbines, etc.) to actively coordinate generation, storage, and consumption, in contrast to traditional systems that simply transmit energy from centralized power plants to users.

7.4.2 The Role of AI in ESSs

In order to maintain a steady power supply, balance generation and demand, and improve grid stability in RESs, energy storage is essential. However, conventional energy storage techniques have a number of drawbacks., including insufficient energy management, safety risks, and variable performance changes. These limitations need for the use of artificial intelligence (AI) in better, more adaptable storage solutions (Fig. 7.10). Predictive maintenance for storage systems, charge-discharge cycle optimization, energy distribution, grid integration, safety improvements, and fault detection are all possible applications for Battery Management Systems (BMS) [4–43].

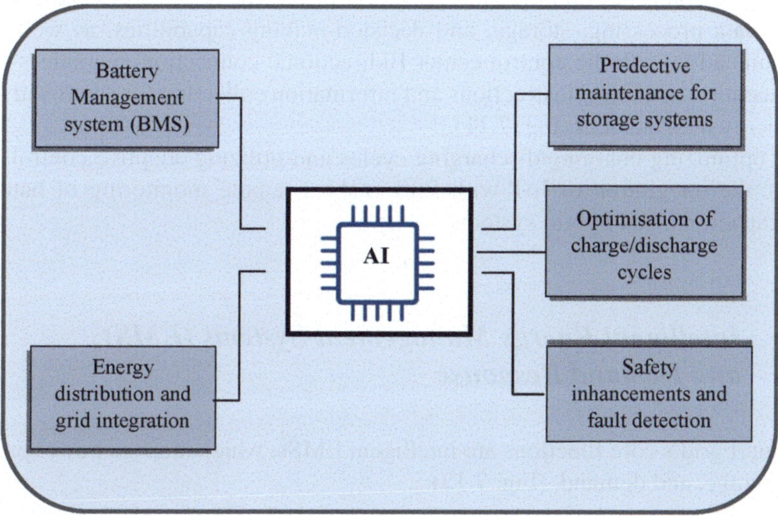

Fig. 7.10 Role of AI in ESSs

7.4.3 Role of AI Within Smart Grids

By predicting the output from sources like wind and solar, AI is essential to the efficient operation of RESs. Regression analysis, neural networks, and time-series forecasting are examples of AI methods that are frequently used for precise energy demand prediction, maximum power point tracking, and design optimization. In smart grids, neural network-based AI methods are particularly popular for predicting future power consumption and assisting in supply and demand balance through constant, real-time load monitoring and management. Furthermore, by optimizing cycles of charging and discharging based on predicted patterns of energy generation and consumption. These features allow AI to significantly increase the stability, and effectiveness.

7.4.4 Communication and Control Architectures Using Internet of Things

The Internet of Things (IoT) makes it possible for sensors, controllers, and smart devices to be easily connected across the grid. IoT-based systems provide continuous data collection on energy production, storage state, consumption trends, and environmental aspects. IoT sensors and controllers that measure real-time parameters (voltage, current, SOC, temperature, etc.) are installed in local equipment like solar panels, wind turbines, monitoring units, and batteries. These devices connect via an IoT gateway or local controller, which uses wireless communication protocols to collect local data and send it to the IoT Cloud Platform. Operators may remotely monitor and optimize the operation of the RESs thanks to the IoT platform's data processing, storage, and decision-making capabilities, as well as its communication with the control center Bidirectional connection guarantees effective execution of control instructions and information collecting for smart and adaptive energy management (Fig. 7.11).

By optimizing charging/discharging cycles and utilizing adaptive control algorithms, the integration of IoT with ESS enables remote monitoring of batteries, supercapacitors, and hybrid systems.

7.4.5 Intelligent Energy Management Systems (EMS) and Demand Response

The smart grid's core functions are intelligent EMSs, which manage power production, storage, and demand. (Fig. 7.12):

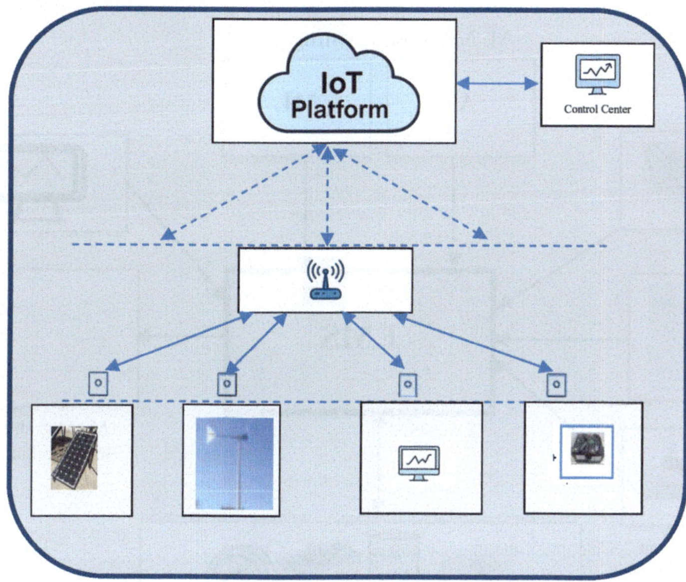

Fig. 7.11 I IoT-Based smart grid with energy storage

- Use machine learning, artificial intelligence, and optimization approaches to make accurate predictions regarding consumption of energy and RE output.
- Storage operations can be efficiently managed by optimizing charge and discharge cycles.
- Promote demand response (DR) programs that allow end users to adjust their energy consumption based on the state of the grid,
- In microgrid and smart community contexts, EMS with DR and ESS optimizes energy costs, increases overall efficiency, and reduces carbon emissions by better balancing supply and demand.

7.5 Future Trends and Research Directions

Digital advancements and developing technologies are driving the next generation of ESSs for the integration of RE. Significant advancements are being made in battery technology, in regards terms of faster charging, and higher energy density. Technologies including hydrogen storage, flow batteries, CAES, supercapacitors, and superconducting are becoming more widely adopted to meet the need for large-scale storage. Digital twins, blockchain-enabled energy transactions, and decentralized energy markets are examples of complimentary digital developments that are making energy ecosystems more transparent, secure, and efficient [44–46].

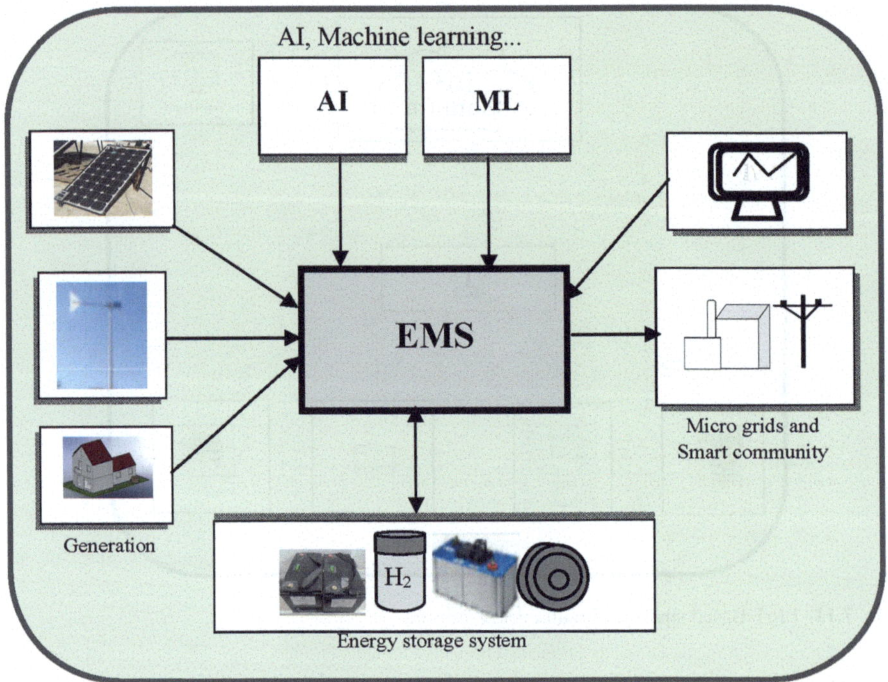

Fig. 7.12 EMS coordinated smart grid

7.6 Conclusion

Because of advancements in heuristic algorithms, mathematical models, and AI-based adaptive control, energy storage optimization in renewable energy systems is always developing. These optimization techniques provide an essential basis for effective energy management, allowing for improved time management, reduced costs and improved system reliability [47–50]. Together, predictive analytics, multi-objective systems, and real-time data offer previously impossible solutions for adjusting energy storage operations to increasing grid demands while reducing environmental impacts and preserving system lifetime. More research into smarter and robust optimization techniques will be essential as the penetration of clean energy sources increases and storage solutions vary.

References

1. Alghamdi AS (2025) Strategic switch placement and renewable wind integration through network reconfiguration with the modified cheetah optimizer. In: IEEE Green Technologies conference (GreenTech), Wichita, USA, pp 123–127. https://doi.org/10.1109/GreenTech62170.2025.10977576
2. Rekioua D (2024) Wind power electric systems: modeling, simulation and control. Green energy and technology. Springer, Cham. https://doi.org/10.1007/978-3-031-52883-5
3. Rekioua D, Matagne E (2012) Optimization of photovoltaic power systems: modeling, simulation and control. Springer. https://doi.org/10.1007/978-1-4471-2403-0
4. Kakouche K, Oubelaid A, Mezani S, Rekioua D, Rekioua T (2023) Different control techniques of permanent magnet synchronous motor with fuzzy logic for electric vehicles: analysis, modelling, and comparison. Energies 16:3116. https://doi.org/10.3390/en16073116
5. Aissou R, Rekioua T, Rekioua D, Tounzi A (2016) Application of nonlinear predictive control for charging the battery using wind energy with permanent magnet synchronous generator. Int J Hydrog Energy 41(45):20964–20973. https://doi.org/10.1016/j.ijhydene.2016.05.249
6. Zhang Z, Lyu Q, Zhang J (2023) Research progress and prospect of distributed PV power generation technology under the goal of emission peak and carbon neutrality. Sol Energy:17–21. https://doi.org/10.19911/j.1003-0417.tyn20211104.01
7. Zhu S, Zhang Y (2024) Coordinated optimization and configuration optimization of wind, photovoltaics and energy storage based on particle swarm optimization algorithm. ICBASE, Wenzhou, pp 450–454. https://doi.org/10.1109/ICBASE63199.2024.10762377
8. Li S et al (2024) An optimization model for joint participation of wind, photovoltaic and storage in the day-ahead electricity spot market considering output uncertainty. EI2, Shenyang, pp 2316–2322. https://doi.org/10.1109/EI264398.2024.10991566
9. Jiang X, Zhang L, Li F, Li Z, Ling Z, Zhao Z (2025) Research on dynamic energy management optimization of park integrated energy system based on deep reinforcement learning. Energies 18:5172. https://doi.org/10.3390/en18195172
10. Altaf A, El Amraoui A, Delmotte F, Lecoutre C (2023) Optimization of temporary storage in cross docking facilities: a mathematical model. In: 7th International Conference on Automation, Control and Robots (ICACR), Kuala Lumpur, Malaysia, pp 147–150. https://doi.org/10.1109/ICACR59381.2023.10314610
11. Yao Y, Zheng F, Wang Z, Yu Y, Zhao X, Ma G (2025) Improved firefly algorithm optimization scheduling technology for grid-connected AC–DC hybrid microgrids. ICoPESA, Nanjing, pp 689–695. https://doi.org/10.1109/ICoPESA65876.2025.11234371
12. Wang L, Wang X, Liu K, Sheng Z (2019) Multi-objective hybrid optimization algorithm using a comprehensive learning strategy for automatic train operation. Energies 12:1882. https://doi.org/10.3390/en12101882
13. Srivastava A, Bajpai R (2019) An efficient maximum power extraction algorithm for wind energy conversion system using model predictive control. IRECON 7(3):93–107. https://doi.org/10.15866/irecon.v7i3.17403
14. Mayilsamy G, Jeong JH, Lee SR, Joo YH (2025) Enhanced active power control with adjustable range of non-pitch regulation for desired reference power tracking in PMVG-based WTS. IEEE Trans Energy Convers 40(2):820–831. https://doi.org/10.1109/TEC.2024.3457516
15. Chen J, Wang G (2025) Stochastic optimal control for maximum energy capture in wind turbines based on dynamic programming principle. CCDC, Xiamen, pp 6640–6645. https://doi.org/10.1109/CCDC65474.2025.11090980
16. Xiong H, Shi Y, Chen Z, Guo C, Ding Y (2023) Multi-stage robust dynamic unit commitment based on fast robust dual dynamic programming. IEEE Trans Power Syst 38(3):2411–2422. https://doi.org/10.1109/TPWRS.2022.3179817
17. Liu C, Liu F, Ma R, Chen Y, Pan L, Xu X (2022) Stochastic scheduling of a wind–photovoltaic–hydro complementary system using stochastic dual dynamic programming. ICSGSC, Chengdu, pp 190–194. https://doi.org/10.1109/ICSGSC56353.2022.9963011

18. Hassan IM, El-Saady G, Ibrahim E, Abdelshafy AM (2022) Optimal sizing of battery/hydrogen renewable energy system with genetic algorithm based on irradiance forecasting using LSTM neural network. MEPCON, Cairo, pp 1–8. https://doi.org/10.1109/MEPCON55441.2022.10021798
19. Zhu S, Zhang Y (2024) Coordinated optimization … (duplicate of #6). ICBASE, Wenzhou, pp 450–454. https://doi.org/10.1109/ICBASE63199.2024.10762377
20. Dong W, Li Y, Xiang J (2016) Sizing of a stand-alone PV/wind energy system with hydrogen and battery storage using improved ant Colony algorithm. CCDC, Yinchuan, pp 4461–4466. https://doi.org/10.1109/CCDC.2016.7531788
21. Mokhtari Y, Rekioua D (2018) High performance of MPPT using ant Colony algorithm in wind turbine. Renew Energy 126:1055–1061. https://doi.org/10.1016/j.renene.2018.03.049
22. Rupa GS, Nuvvula RSS, Kumar PP, Ali A, Khan B (2024) Machine learning-based optimization techniques for renewable energy systems. icSmartGrid, Setubal, pp 389–394. https://doi.org/10.1109/icSmartGrid61824.2024.10578295
23. Huang S, Li P, Yang M, Gao Y, Yun J, Zhang C (2021) A control strategy based on deep reinforcement learning for wind–solar storage systems. IEEE Trans Ind Appl 57:6547–6558. https://doi.org/10.1109/TIA.2021.3105497
24. Buasri P, Salameh ZM (2007) Modeling of a distributed generation system using ANFIS. In: IEEE PES general meeting, Tampa, pp 1–9. https://doi.org/10.1109/PES.2007.386000
25. Yang J (2024) Deep learning methods for smart grid data analysis. SGEI, Shenyang, pp 717–720. https://doi.org/10.1109/SGEI63936.2024.10914173
26. Barakat S, Mageed HMA, Samy M (2023) Eco-efficient mobility: comparative optimization of PV–wind EV charging solutions. MEPCON, Mansoura, pp 1–7. https://doi.org/10.1109/MEPCON58725.2023.10462329
27. Hassani H, Zaouche F, Rekioua D, Belaid S, Rekioua T, Bacha S (2020) Feasibility of a stand-alone PV/battery system with hydrogen production. J Energy Storage 31:101644. https://doi.org/10.1016/j.est.2020.101644
28. Zaouche F, Rekioua D, Gaubert JP, Mokrani Z (2017) Supervision and control strategy for PV generators with battery storage. Int J Hydrog Energy 42:19536–19555. https://doi.org/10.1016/j.ijhydene.2017.06.107
29. Rekioua D, Rekioua T, Soufi Y (2015) Control of a grid-connected photovoltaic system. ICRERA, Palermo, pp 1382–1387. https://doi.org/10.1109/ICRERA.2015.7418634
30. Amereh M, Khozani ZS, Kazemi A (2014) Multi objective design of stand-alone PV/wind system using hybrid GA–PSO. ICEE, Tehran, pp 695–699. https://doi.org/10.1109/IranianCEE.2014.6999628
31. Ashkezari LS, Mouli GRC, Yorke-Smith N, Bauer P (2024) Multiobjective system sizing for heavy-duty EV charging stations. ESARS-ITEC, Naples, pp 1–6. https://doi.org/10.1109/ESARS-ITEC60450.2024.10819791
32. Fesseler A, Rothmund C, Schmidt L, Bank D, Kallo J (2026) Investigation of a cost-effective energy-storage system in the DC bus. IEEE Trans Power Electron 41(2):2491–2506. https://doi.org/10.1109/TPEL.2025.3605446
33. Sterner M, Stadler I (2019) Handbook of energy storage. Springer
34. Wang K, Zhou S, Lu S, Li Z (2022) Economic analysis of energy storage power stations in distribution networks. IC2ECS, Nanjing, pp 187–191. https://doi.org/10.1109/IC2ECS57645.2022.10088061
35. Han Z, Chi R, Wang N (2025) Research on economic analysis and comprehensive evaluation method of compressed air energy storage. In: International conference on electrical automation and artificial intelligence (ICEAAI), Guangzhou, China, pp 278–283. https://doi.org/10.1109/ICEAAI64185.2025.10956448
36. Franey JP, Okrasinski TA, Schaeffer WJ (2010) Reducing environmental impact through packaging: a life-cycle assessment. Bell Labs Tech J 15(2):193–204. https://doi.org/10.1002/bltj.20449

37. Balakrishnan A et al (2019) Environmental impacts of utility-scale battery storage in California. PVSC, Chicago, pp 2472–2474. https://doi.org/10.1109/PVSC40753.2019.8980665
38. Zhang J, Basson L, Leach M (2009) Review of LCA studies of coal-fired power plants with CCS. SUPERGEN, Nanjing, pp 1–7. https://doi.org/10.1109/SUPERGEN.2009.5348020
39. Refaat SS, Ellabban O, Bayhan S, Abu-Rub H, Blaabjerg F, Begovic MM (2021) Smart grid architecture overview. In: Smart grid and enabling technologies. IEEE, pp 1–43. https://doi.org/10.1002/9781119422464.ch1
40. Gungor V et al (2011) Smart grid & IoT: communication & control. IEEE Trans Ind Informatics
41. Pradeep S, Krishna S, Reddy MS, Sri DD, Sri MS (2023) Analysis and functioning of smart grid for enhancing energy efficiency using IoT. ICCCMLA, Hamburg, pp 316–321. https://doi.org/10.1109/ICCCMLA58983.2023.10346767
42. Han X, Li K (2024) AI-orchestrated data-driven modelling and control in BMS. IAI, Shenyang, pp 1–7. https://doi.org/10.1109/IAI63275.2024.10730148
43. Ivanov G, Mihaylova EM, Mihalevska N (2025) Self-adaptive algorithms for control of hybrid energy systems with AI elements. COMSCI, Sozopol, pp 1–4. https://doi.org/10.1109/COMSCI67172.2025.11225183
44. Wu J (2025) AI optimization for coordinated scheduling of distributed energy storage. AESPE, Hangzhou, pp 17–21. https://doi.org/10.1109/AESPE65805.2025.11162913
45. Panwar D, Mishra SS, Shanmugapriya V (2024) AI status, prospects and challenges in smart grids for hydrogen energy. Global AI Summit, India, pp 1065–1070. https://doi.org/10.1109/GlobalAISummit62156.2024.10947929
46. Ba L, Tangour F, El Abbassi I, Absi R (2025) Analysis of digital twin applications in energy efficiency: a systematic review. Sustainability 17:3560. https://doi.org/10.3390/su17083560
47. Gariya N, Asrani A, Mandal A, Shaikh A, Cha D (2025) Integrating blockchain into the energy supply chain for transparency and sustainability. Energies 18:2951. https://doi.org/10.3390/en18112951
48. Rekioua D, Rekioua T, Elsanabary A, Mekhilef S (2023) Power management of an autonomous PV/wind/battery system. Energies 16:2286. https://doi.org/10.3390/en16052286
49. Rekioua D, Roumila Z, Rekioua T (2008) Étude d'une Centrale Hybride Photovoltaïque–Éolien–Diesel. J Renew Energ 11(4):623–633. https://doi.org/10.54966/jreen.v11i4.112
50. Rekioua D (2019) Hybrid renewable energy systems: optimization and power management control. Green energy and technology. Springer, Cham. https://doi.org/10.1007/978-3-030-34021-6

Chapter 8
Applications of Energy Storage

8.1 Introduction

This chapter presents a series of applications of hybrid RESs with various generation sources and storage technologies. It illustrates how PV panels, wind turbines, batteries, fuel cells (FCs), and supercapacitors (SCs) can be combined to achieve reliable, autonomous, and efficient energy systems across different contexts. Each section explores a specific configuration, from off-grid PV-battery systems to wind turbine/flywheel hybrids, and from PV/fuel cell microgrids to PEMFC/SC systems for electric vehicles, highlighting their operating principles, control strategies, and simulation results under various conditions. The aim is to connect theoretical modeling with practical applications. Through comparative case studies, this chapter gives the significance of storage coordination, energy flow optimization, and multi-source integration in sustainable power systems. These examples are based on my own experience and on my previous works [1–24].

8.2 Application of Chap. 3

8.2.1 Application of PV System with Batteries Storage in Remote Area

To analyse how battery storage impacts the ability of a PV system to meet a constant daily energy demand in a remote area, we take an application of constant load with a daily energy demand of $E_{daily} = 10$ kWh supplied by a PV generator with a peak power of $P_{PV,peak} = 5$ kW$_P$. The solar generation hours are 6 h/day and we take a as assumption battery storage capacity $E_{batt,max} = 7$ kWh. We want to analyze the

D. Rekioua, *Energy Storage for Renewable Energy Systems*, Green Energy and Technology, https://doi.org/10.1007/978-3-032-19589-0_8

system's performance, energy generation, and consumption over a single day (from hour 0 to hour 24) for two scenarios: with and without batteries.

Scenario1—without battery storage

- PV can supply load only during daylight.
- Total PV energy supplied in 24 h: $E_{PV} = 2.5$ kWh
- Available energy: $E_{available} = E_{daily} - E_{PV} = 10–2.5 = 7.5$ kWh
- the percentage of the demand met P_{cov} is:

Scenario 2—with battery Storage

- Daytime: excess PV production charges the battery.
- Night time: battery discharges to supply the load.
- Total energy supplied (PV + battery):

$$E_{supplied} = 7.5kWh$$

And if we want 75% coverage (7.5 kWh/day), the battery must supply:

$$E_{batt-needed} = 7.5 - 2.5 = 5kWh / day$$

So, the required battery usable energy must be at least:

$$E_{batt-usuable} \geq 5kWh / day$$

To achieve 100% coverage, we can either:

- Increase PV capacity to produce more during daylight hours,
- Increase the number of solar hours (through tracking systems),
- Reduce the daily load,
- Or combine PV with another renewable source (as wind).

We can calculate the hourly energy demand as:

$$HED = \frac{E_{daily}}{24\ hours} \quad (8.2)$$

$$HED = 0.42kWh / hour$$

This means the user consumes 0.42 kWh every hour evenly throughout the day.

The solar production depends on the PV system power rating, the effective solar irradiance in a given hour, and the system performance:

$$E_{PV} = P_{PV} \cdot I_h \cdot \eta_{losses} \quad (8.3)$$

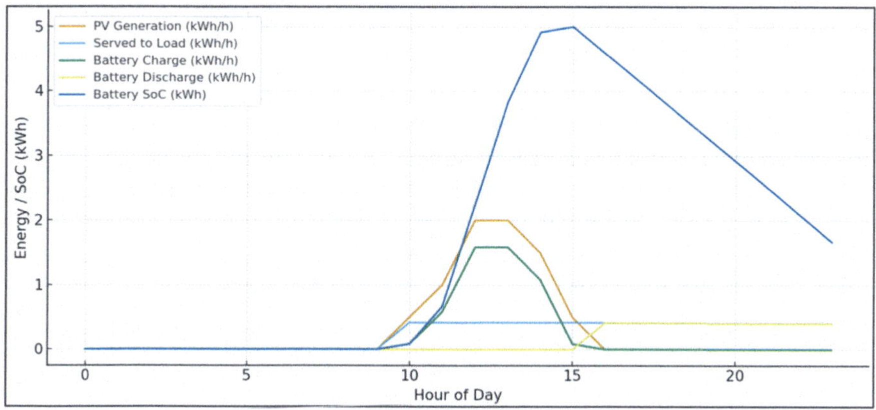

(a) PV battery flows and state of charge during 24 hours

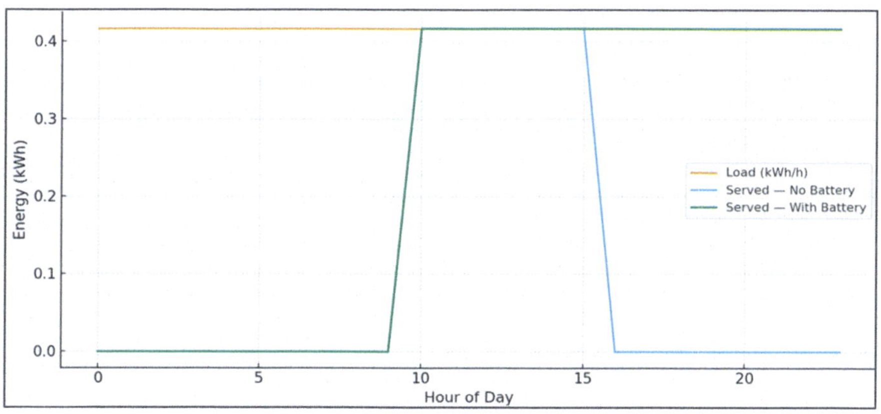

(b) Saved energy vs load (with and without battery)

Fig. 8.1 Variations of the various energies in a PV system. (**a**) PV battery flows and state of charge during 24 h. (**b**) Saved energy vs load (with and without battery)

The different energies from solar production, daily demand, solar to demand, solar excess to battery and available one are represented in Fig. 8.1. The data make it evident how a PV system's utility under continuous load is altered by battery storage. Seventy-five percent of the demand is unmet without storage, as the majority of solar power is squandered due to its limited six-hour window. By shifting excess midday energy into the evening and night, a battery can increase coverage from 25% to more than half of the daily load in some situations and to 75% in steady-state operation, when afternoon surplus energy can be used the next morning.

The total daily demand is about 10 kWh. Without a battery, there will be an unmet demand of 0.42 kWh × 12 night hours = 5.04 kWh. With a battery, unmet demand only occurs before solar production starts (the first 6 h), resulting in a shortfall of 0.42 kWh × 6 = 2.52 kWh.

8.2.2 Application of PV/Battery Storage in Electric Vehicle Charging

PV systems with battery storage are increasingly integrated into electric vehicle (EV) applications to enhance energy autonomy and reduce dependence on grid charging. Also, standalone solar charging stations with battery storage allow EV charging in remote or off-grid areas, ensuring continuous operation even in cloudy conditions. Figure 8.2 illustrates this application.

We can have two scenarios:

- Scenario 1 (Direct Charging of the EV and battery charge): The PV-generated power is sent to the EV's on board charger to recharge its battery when solar irradiance is available.

$$\text{If } P_{PV}(t) \geq P_{EV-req}(t) \tag{8.5}$$

- Then, any excess PV energy not immediately used by the EV is stored in the stationary battery storage unit. To ensure charging availability regardless of solar conditions, the EV can then be charged at night or during cloudy conditions using this stored energy.
- Scenario 2 (Battery discharges to supply EV): If the demand for EV charging power exceeds the PV power:

$$\text{If } P_{PV}(t) < P_{EV-req}(t) \tag{8.6}$$

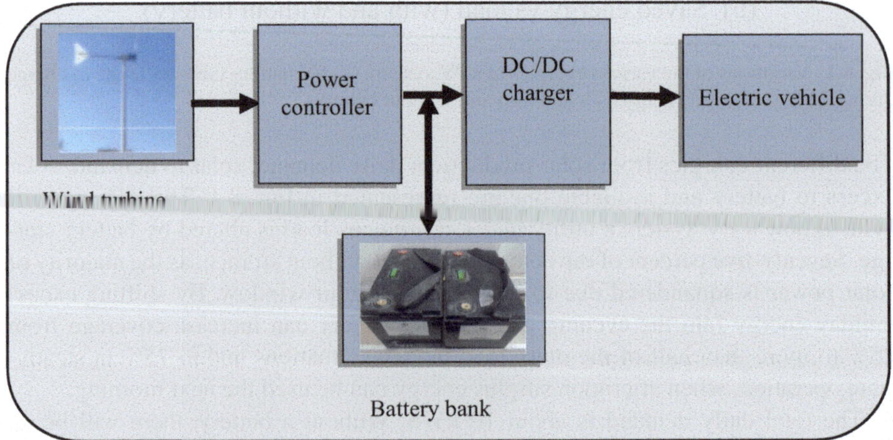

Fig. 8.2 PV/battery storage in electric vehicle for off-grid EV charging

- Then, the battery discharges to supplement the deficit, ensuring continuous EV charging. This operation guarantees that EV charging can be sustained even without immediate solar input.

A battery with a nominal capacity *of* $E_{batt,nom}$ = 10 kWh is used to support the charging of an electric vehicle in an isolated solar installation.

- During a controlled test, the battery is charged from 20% *SOC* to 80% *SOC*, requiring a measured electrical energy input from the charger of $E_{supplied}$ = 6.40 kWh.
- In a second test, the battery is discharged from 80% *SOC* back to 20% SOC into a resistive load, and the energy delivered to the load is measured as:

$$E_{delivered} = 5.85\,kWh$$

- The stored energy is the amount of energy currently inside the battery at a given time:

$$E_{stored} = \Delta SOC.E_{batt-nom} \tag{8.7}$$

$$E_{stored} = 0.6 \times 10 = 6.0\,kWh$$

The removed energy is the amount of energy taken out of the battery during a discharge cycle. It is given by:

$$E_{removed} = \Delta OC\,.\,E_{batt-nom} \tag{8.8}$$

$$E_{removed} = 0.6 \times 10 = 6.0kWh$$

So, the charging efficiency is:

$$\eta_{charge} = \frac{E_{stored}}{E_{supplied}} \tag{8.9}$$

$$\eta_{charge} = \frac{6.0}{6.40} = 9375 = 93.75\%$$

The discharging efficiency is:

$$\eta_{discharge} = \frac{E_{delivered}}{E_{removed}} \tag{8.10}$$

$$\eta_{discharge} = \frac{5.85}{6} = 0.975 = 97.5\%$$

The round-trip efficiency (RTE) is:

$$RTE = \frac{E_{delivered}}{E_{supplied}} \tag{8.11}$$

$$RTE = \frac{5.85}{6} = 0.9140 = 91.40\%$$

$$RTE = \eta_{charge} \cdot \eta_{discharge} \tag{8.12}$$

$$RTE = 0.9375 \times 0.975 = 0.9140 = 91.40\%$$

An RTE of 91.4% means that for every 1 kWh supplied to the battery, you get back 0.914 kWh after charging and discharging.

8.2.3 Application of PV/FCs in Electric Vehicle

An electric vehicle (EV) uses a PV array and a fuel-cell system. The PV array has a rated power of 3 kW and operates for 5 h/day at peak output, while the fuel cell provides a constant power of 2 kW when active. The EV's daily energy consumption is 18 kWh.

If the PV system cannot meet the daily demand, the fuel cell runs to supply the extra energy. Calculate:

1. The daily energy supplied by the PV system.

$$E_{PV} = P_{PV} \cdot t \tag{8.13}$$

$$E_{PV} = 3 \times 5 = 15\, kWh / day$$

2. The remaining energy to be supplied by the fuel cell.

$$E_{FC} = E_{consumption} - E_{PV} \tag{8.14}$$

$$E_{FC} = 18 - 15 = 3\, kWh / day$$

3. The number of operating hours per day required for the fuel cell

$$t_{FC} = \frac{E_{FC}}{P_{FC}} \tag{8.15}$$

$$t_{FC} = \frac{3}{2} = 1.5\,\text{h} / day$$

4. If the hydrogen lower heating value (LHV) is 33.3 kWh/kg and fuel-cell system (tank-to-bus) electrical efficiency (η_{FC}) about 50%, calculate hydrogen mass required per day.

$$m_{H_2} = \frac{E_{FC}}{P_{FC} \cdot \eta_{FC}} \tag{8.16}$$

$$m_{H_2} = \frac{3}{33.3 \times 0.5} = 0.180\,kg$$

At 50% system efficiency the vehicle needs only 0.18 kg H_2 per day to cover the 3.0 kWh. It is a small mass because hydrogen has very high gravimetric energy density. The usable electrical energy via a fuel cell at 50% efficiency is the *effective* gravimetric energy density in terms of electricity delivered to the load:

$$P_{FC} \cdot \eta_{FC} = 33.3 \times 0.5 = 16.65\,kWh / kg$$

5. For 7 days' autonomy, the required hydrogen mass is $m_{H2\text{-}7days}$ = 0.18 kg/day × 7 days = 1.26 kg. If we add design reserve, it is good practice to add a safety/usable-reserve margin, a common choice is ≈20%margin. We will have :

$$m_{H_{2-margin}} = \frac{m_{H_{2-7\,days}}}{(1 - 0.2)} \tag{8.17}$$

$$m_{H_{2-margin}} = \frac{1.26}{0.8} = 1.58\,kg$$

To convert hydrogen mass to tank volume, we know that volume depends on storage pressure and temperature, for example a pressure of 350 bar, we have$\rho \approx 23$ kg/m^{3}and for 700 bar, we have $\rho \approx 42$ kg/m^3.
So, the volume required at 350 bar is:

$$\text{V} = 1.58\,kg / 23\,kg / \text{m}^3 \approx 0.069\,\text{m}^3 = 69\,\text{L},$$

and at 700 bar, the volume is.

$$V = 1.58\,kg / 42\,kg / m^3 \approx 0.038\,m^3 = 38\,L.$$

Thus a small cylinder of order ~[40–70 l] internal volume (depending on pressure [350–700 bars]) is enough for ~1.58 kg H_2.

8.2.4 Application in Combinations

8.2.4.1 Example 1

If we consider just the ($n_c = 11$) most common storage technologies, we will have:

$$C = 2^{11} - 11 - 1 = 2048 - 12 = 2036$$

So with 11 technologies, you can obtain 2036 possible hybrid storage configurations (Table 8.1).

And we want all hybrid combinations of size $k_c = 2$ and $k_c = 3$, with **no** singles ($k_c \neq 1$), no repetition of the same technology in a pair (combinations without replacement) and the order does not matter (BES + SES=SES + BES), the counts follow standard combinations are:

$$Number \quad of \quad pairs = \binom{n_c}{2} \tag{8.18}$$

$$Number \quad of \quad pairs = \binom{11}{2} = 55$$

Table 8.1 Number of combinations for each subset size

Subset size k_c	Name	Number of combinations (11 k)
0	Empty set	1
1	Singles	11
2	Pairs	55
3	Triplets	165
4	Quartets	330
5	Quintets	462
6	Sextets	462
7	Septets	330
8	Octets	165
9	Nonets	55
10	Decades	11
11	Whole set	1

$$Number \quad of \quad triples = \binom{n_c}{3} \tag{8.19}$$

$$Number \quad of \quad triples = \binom{11}{3} = 165$$

$$Number \quad of \quad configurations = \binom{n_c}{2} + \binom{n_c}{3} \tag{8.20}$$

$$Number \quad of \quad configurations = 55 + 165 = 210$$

8.2.4.2 Example 2

Suppose we have 4 storage technologies characterized by their cost and efficiency (Table 8.2):

And we would like to calculate the objective function for all combination (Table 8.3).

where w_c and w_e are weights applied to cost and efficiency, respectively.

- Let $w_c = -1$: this means we want to minimize total cost because multiplying cost by -1 makes lower cost increase the objective value (if we are maximizing f(c)).
- Let $w_e = +5$: this means we want to prioritize efficiency strongly by giving it a positive and larger weight, emphasizing the importance of high efficiency in the overall score.

$$f(c) = (-1).(cost \quad sum) + (5).(cost \quad efficiency) \tag{8.21}$$

- The least negative (best) value is for pair $\{s_1, s_3\}$ with $f(c) = -13.4$. To obtain positive f(c) values, let's adjust the weights from $w_c = -1$ and $w_e = +5$ to $w_c = -0.5$, $w_e = +10$ (Table 8.4).

Under these weights, all pairs now have positive f(c)values and the best pair according to f(c) is $\{s_1, s_3\}$ with $f(c) = 6.2$.

Table 8.2 Example of costs and efficiencies of Si technologies

Technology	Cost (c_i)	Efficiency (e_i)
s_1	10	0.85
s_2	15	0.90
s_3	12	0.87
s_4	20	0.95

Table 8.3 Example 1 of calculations of pairs of multi-storages technologies

Pair	Cost sum	Efficiency sum	f(c) calculation	f(c) value
s_1,s_2	10 + 15 = 25	0.85 + 0.90 = 1.75	−1 × 25 + 5 × 1.75 = −25 + 8.75–1 × 25 + 5 × 1.75 = −25 + 8.75	−16.25
s_1,s_3	10 + 12 = 22	0.85 + 0.87 = 1.72	−1 × 22 + 5 × 1.72 = −22 + 8.6–1 × 22 + 5 × 1.72 = −22 + 8.6	−13.40
s_1,s_4	10 + 20 = 30	0.85 + 0.95 = 1.80	−1 × 30 + 5 × 1.80 = −30 + 9–1 × 30 + 5 × 1.80 = −30 + 9	−21.00
s_2,s_3	15 + 12 = 27	0.90 + 0.87 = 1.77	−1 × 27 + 5 × 1.77 = −27 + 8.85–1 × 27 + 5 × 1.77 = −27 + 8.85	−18.15
s_2,s_4	15 + 20 = 35	0.90 + 0.95 = 1.85	−1 × 35 + 5 × 1.85 = −35 + 9.25–1 × 35 + 5 × 1.85 = −35 + 9.25	−25.75
s_3,s_4	12 + 20 = 32	0.87 + 0.95 = 1.82	−1 × 32 + 5 × 1.82 = −32 + 9.1–1 × 32 + 5 × 1.82 = −32 + 9.1	−22.90

Table 8.4 Example 2 of calculations of pairs of multi-storages technologies

Pair	Cost sum	Efficiency sum	f(c) = −0.5 × Cost Sum + 10 × Efficiency Sum	f(c) value
s_1,s_2	25	1.75	−0.5 × 25 + 10 × 1.75 = −12.5 + 17.5–0.5 × 25 + 10 × 1.75 = −12.5 + 17.5	5.0
s_1,s_3	22	1.72	−0.5 × 22 + 10 × 1.72 = −11 + 17.2–0.5 × 22 + 10 × 1.72 = −11 + 17.2	6.2
s_1,s_4	30	1.80	−0.5 × 30 + 10 × 1.80 = −15 + 18–0.5 × 30 + 10 × 1.80 = −15 + 18	3.0
s_2,s_3	27	1.77	−0.5 × 27 + 10 × 1.77 = −13.5 + 17.7–0.5 × 27 + 10 × 1.77 = −13.5 + 17.7	4.2
s_2,s_4	35	1.85	−0.5 × 35 + 10 × 1.85 = −17.5 + 18.5–0.5 × 35 + 10 × 1.85 = −17.5 + 18.5	1.0
s_3,s_4	32	1.82	−0.5 × 32 + 10 × 1.82 = −16 + 18.2–0.5 × 32 + 10 × 1.82 = −16 + 18.2	2.2

8.2.5 Application of a Wind Turbine/Battery Storage

The studied system consists of the following components: a 1 kW PMSG wind turbine that converts wind energy into electrical energy, a diode bridge rectifier, a DC-DC boost converter controlled by MPPT algorithm, a battery bank connected via a bidirectional converter, a single-phase voltage source inverter (VSI) and control strategies used to regulate voltage, current, frequency, and power quality fed into the grid or autonomous load (Fig. 8.3):

8.2.5.1 Sizing of the Studied System

For the sizing, we have to calculate the sizes and ratings of the main components (turbine rotor, rectifier, boost converter, battery, bidirectional converter, inverter) from (Table 8.5):

The Inputs required from the user are (Table 8.6):

8.2.5.1.1 Wind Turbine Sizing

We size rotor/diameter for rated power at a reference speed $V_{wind\text{-}ref}$: [13–18].

$$\begin{cases} A_{Tb} = \dfrac{2.P_{Tb}}{C_p.\varphi_{air}.V_{wind-ref}^3} \\ D = \sqrt{\dfrac{4A}{\pi}} \end{cases} \tag{8.22}$$

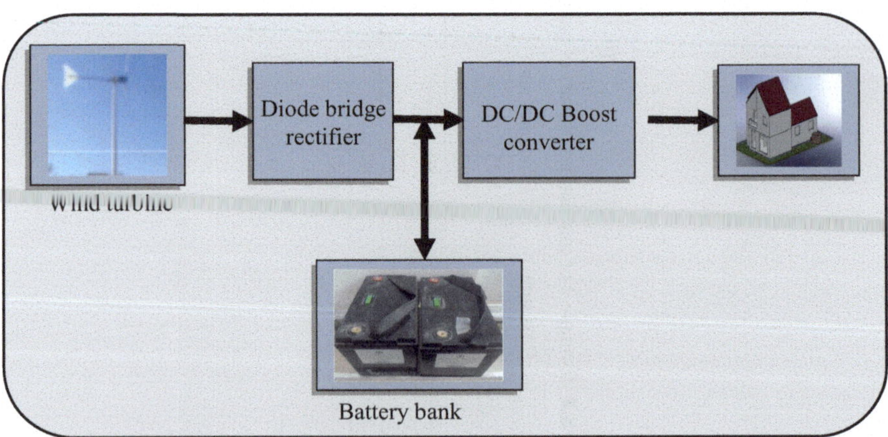

Fig. 8.3 Studied system

Table 8.5 Sizes of the main components

Site characteristics	Wind resource: wind speed distribution
Load needs	Load profile: Hourly power/daily energy, desired autonomy
Component parameters	Efficiencies, Battery *DOD*, Nominal dc voltage, Safety margin

Table 8.6 Required inputs

Inputs	Description
Wind data	Average speed (m/s) or Weibull distribution (k, c), Analysis period (day/month/year
Turbine characteristics:	Rated power (here 1 kW), $C_{p\text{-}max}$ (aerodynamic efficiency ≈ 0.35–0.45), Air density (ρ_{air} = 1.225 kg/m^3), Nominal speed
Load profile	Hourly curve (kW), Or daily energy consumption (kWh)
Battery parameters	Nominal voltage (Vdc), Usable dod (e.g. 0.8), Round-trip efficiency (≈0.85–0.95).
Converter parameters	Efficiencies (rectifier, boost, bidirectional, inverter), safety margin (%) for current/power.
Desired autonomy	Days and confidence level (% of hours covered by wind if using distribution)

The obtained findings are given in Table 8.7.

8.2.5.1.2 Battery Sizing

The nominal battery capacity is calculated by using the required stored energy [1–4].

$$\begin{cases} E_{batt-req} = E_{load,day} \cdot N_{aut} \\ C_{batt} = \dfrac{E_{batt-req}}{DOD.\eta_{batt} \cdot V_{batt}} \end{cases} \tag{8.23}$$

The obtained findings are shown in Table 8.8.

The selected battery will be ≈ 48 V, 370 Ah.

8.2.5.1.3 Boost and Bus Sizing

The design of the current I_{DC} is made using these equations with a margin of 25%

Table 8.7 Findings of wind turbine sizing

Inputs		Results	
Rated power (kW)	1	Rotor area A_{Tb}(m²)	2,36
Air density (kg/m³)	1225	Rotor diameter D (m)	1,73
C_p (power coefficient)	0,4		
Reference wind speed $V_{wind\text{-}ref}$ (m/s)	12		

Table 8.8 Findings of battery sizing

Inputs		Results	
$E_{load,\ day}$ (kWh)	6	Nominal energy required $E_{batt-req}$ (kWh)	12
N_{aut} (days)	2		
V_{batt} (V)	48	Battery capacity C_{batt} (Ah)	367.64
DOD	0.8		
η_{batt}	0.85		

$$\begin{cases} I_{DC} = \dfrac{P_{\max}}{V_{DC}} \\ I_{DC-design} = I_{DC} \times 1.25 \end{cases} \tag{8.24}$$

The obtained findings are shown in Table 8.9.

The chosen sizing current I_{DC} is ≈26A, so for the inverter, we choose $P_{nom} \geq 1.2$ kW.

8.2.5.2 Modelling of the Studied System

Table 8.10 lists the many equations used to model the system under study [2, 5, 6, 19–24]:

8.2.6 *Application of Wind Turbine/FESS*

In this application, we consider a WTb based on self-excited induction generator with FESS where the exact parameters embedded and controller blocks are shown in Fig. 8.4.

A 7.5 kW WTb connected via a gearbox to a 5.5 kW self-excited) induction generator (SEIG) makes up the mechanical subsystem. The gearbox converts the turbine's low-speed, high-torque revolution into the higher-speed operation needed by the generator. The same shaft is coupled to a flywheel storage system, which provides short-term kinetic energy storage to reduce torque variations caused on by wind fluctuations.

Table 8.9 DC bus sizing

Inputs		Results	
DC voltage (V)	48	DC current (A)	20.83
Rated power (kW)	1	Design DC current with margin (A)	26.04

Table 8.10 Various equations used to simulate Wind turbine with batteries

Wind turbine	PMSG	
$P_{Tb} = \left(\frac{1}{2}\right).C_p.\rho.A_{Tb}.V_{wind}^3$ $A_{Tb} = \pi.D^2/4$ $T_{Tb} = \left(\frac{1}{2}\right).C_p.\rho.\pi.R_{Tb}^5.\frac{\omega_{Tb}^2}{\lambda_{opt}^3}$ $\lambda_{opt} = \omega_{Tb-opt}.R_{Tb}/V_{wind}$	$V_{sd} = R_{stator}.I_{sd} + L_d\left(\frac{dI_{sd}}{dt}\right) - L_q.\omega.I_{sq}$ $V_{sq} = R_{stator}.I_{sq} + L_q\left(dI_{sq}/dt\right) + L_d.\omega.I_{sd} + \Phi_f.\omega$ $\omega = P.\Omega$ $T_{em} = \left(\frac{3}{2}\right)\left[\Phi_f.I_{sq} + \left(L_d - L_q\right).I_{sd}.I_{sq}\right]$ $J.(d\omega_{Tb}/dt) = T_{Tb} - T_{em} - f.\omega_{Tb}$ $J = J_{Tb} + G^2.J_g$	
Puissance DC après redressement et boost	Battery	
$P_{DC} = P_{Tb}.\eta_{gen}.\eta_{rect}.\eta_{boost}$	$V_{batt} = E_0 - R_{Batt}.I_{batt} - k.\int\left(\frac{I_{batt}}{Q}\right).dt$ $SOC = 1 - \frac{I_{batt}.t}{C_{batt}}$	

On the electrical side, a diode bridge rectifies the generator output to supply the DC bus, which uses a DC-link capacitor to maintain a controlled voltage (usually approximately 450 V). The two primary PI regulators in the control structure are one for the induction generator's flux and torque control ($K_{p1} = 17$), as well as one for DC-bus voltage control ($K_{p2} = 0.908$). These controllers provide steady DC voltage, steady generator operation, and efficient energy transmission between the turbine, generator, and storage system. Smooth power supply under variable wind conditions is made possible by the flywheel's mechanical inertia and the coordinated control loops.

8.2.6.1 Studied System Sizing

- **Steps to size:**

Table 8.11 provides an overview of the sizing steps:

The different input parameters used in this application have been used in equations to determinate the sizing of each subsystem [24] (Table 8.12).

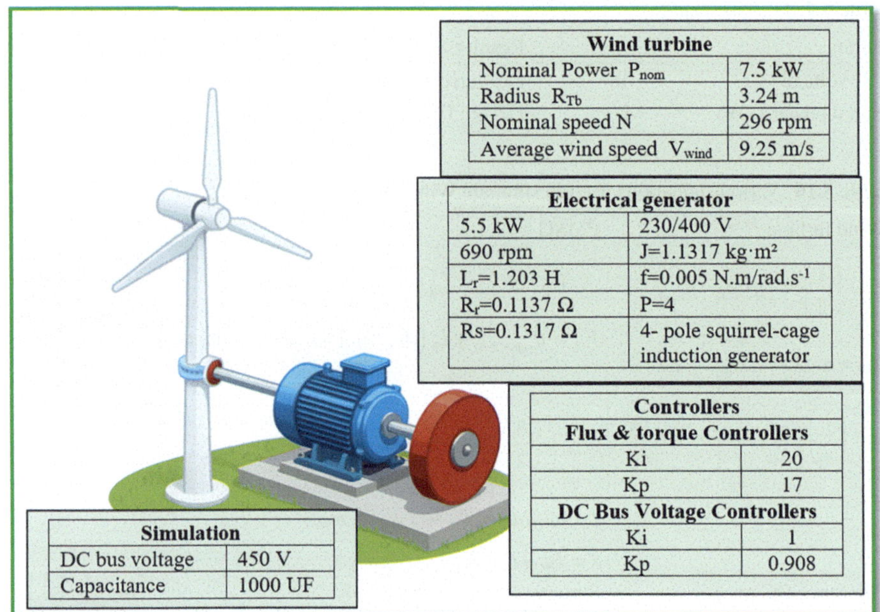

Fig. 8.4 Wind turbine system with induction generator and flywheel energy storage

Table 8.11 Sizing steps

Parameter to size	What to do
Wind turbine	Calculate rotor area from the radius; Use the wind speed and power equation to estimate the power output.
Induction generator	Select power rating, Voltage, Poles (we have 4 pairs), Resistances, and reactances
Flywheel storage	Determine energy storage capacity using moment of inertia and angular velocity; Size mass and dimensions according to material strength.
Control systems	Tune PI controllers for flux, torque, and voltage with provided gains

8.2.6.2 Modelling of the Studied System

Table 8.13 provides a summary of the modeling steps [13–24].

Table 8.12 Sizing of each subsystem based on the provided parameters

Subsystem	Input parameters	Equations	Results
Wind turbine rotor	Radius $R_{Tb} = 3.24$ m	Swept area: $A_{Tb} = \Pi . R_{Tb}^2$	$A_{Tb} \approx 32.98$ m^2
	Wind speed $V_{wind} = 9.25$ m/s	Wind power: $P_{wind} = \left(\frac{1}{2}\right) \mathrm{P}_{air} . A_{tb} . V_{wind}^3$	$P_{wind} \approx 16.0$ kW
	Air density $\rho_{air} = 1.225$ kg/m^3 $C_{p\text{-}max} = 0.593$	Maximum extractable power: $P_{Tb} = \left(\frac{1}{2}\right) . C_{P\text{-}max} . \mathrm{P}_{Air} . A_{Tb} . V_{Wind}^3$	$P_{max} \approx 9.5$ kW
	Rated power $P_{Tb} = 7.5$ kW Rated speed $\Omega_{Tb} = 296$ rpm	Tip-speed ratio: $\lambda = \frac{\Omega_{Tb} . R_{Tb}}{V_{Wind}} = \frac{\Omega_{Tb} . \left(\frac{2\pi}{60}\right) . R_{Tb}}{V_{Wind}}$	$\lambda \approx 10.85$
Induction generator	Rated power $P_{IG} = 7.5$ kW	Synchronous speed: $N_s = \frac{60 . f}{p}$	$N_s = 750$ rpm
	Pole pairs $P = 4$	Slip: $s = \frac{N_s - N_r}{N_s}$	$s = 0.08$
	Rated speed $N_r = 690$ rpm	Mechanical speed: $\Omega_{mec} = \frac{2.\Pi . N_R}{60}$	Ω_{mec} $= 72.26$ rad/s
	Stator/rotor resistances $R_s = 1.07131\ \Omega$, $R_r = 1.29511\ \Omega$	Electromagnetic torque: $T_{em} = \frac{P_{tb}}{\Omega_{mec}}$	$T_{em} = 103.75$ N.m
	Stator/rotor inductances $L_s = 0.1137$ mH, $L_r = 0.1096$ mH	Equivalent circuit used for current and torque modeling	Parameters for simulation
Flywheel storage	Speed $N = 296$ rpm	Angular velocity: $\omega = \frac{2.\Pi . N}{60}$	$\omega = 31{,}0$ rad/s
	Moment of inertia $J = 0.230$ kg·m^2	Energy: $E = \frac{1}{2} . J_{inertia} . \omega^2$	$E = 110.51$ joules
Control system PI Flux/Torque Kp=17, Ki=20 PI Vdc Kp=0.908, Ki=1	PI gains flux/torque $K_I = 20$, $k_p = 17$	Pi controller output: $U(t) = K_p . e(t) + K_I \int E(t) . dt$	Control gains for stability
	PI gains voltage: $K_I = 1$, $k_p = 0.90$	Separate loops for flux, torque, and voltage	Regulator parameters

Table 8.13 Modelling of the studied system

Sub-system	Main variables	Equations
Aerodynamic rotor	Wind speed (V_{Wind}) Rotor speed (Ω_{Tb}) Tip–speed ratio (λ) Power coefficient ($C_P(\lambda,\beta)$) Torque (T_{Tb})	$P_{Tb} = \left(\frac{1}{2}\right) . C_P . \mathrm{P}_{Air} . A_{Tb} . V_{Wind}^3$ $A_{Tb} = \Pi . R_{Tb}^2$ $\lambda = \frac{\omega_{Tb} . R_{Tb}}{V_{wind}}$ $T_{Tb} = \left(\frac{1}{2}\right) . C_p . \rho . \pi . R_{Tb}^5 . \frac{\omega_{Tb}^2}{\lambda_{opt}^3}$
Shaft + Gearbox	Generator speed (ω_g) Electromagnetic torque (T_{em}) Mechanical shaft torque at the generator side (T_g)	$\omega_g = G.\ \omega_{Tb}$ $T_{em} = \frac{T_{Tb}}{G}$ $J_{inertia} . \left(\frac{d\omega_g}{dt}\right) = T_g - T_{em} - f.\omega_g$ $J_{inertia} = J_g + \frac{J_{Tb}}{G^2}$
SCIG (dq, stator frame)	Stator/rotor currents (I_{sd}, I_{sq}, I_{rd}, I_{sq}) Stator fluxes (ϕ_{sd}, ϕ_{sq}, ϕ_{sd}, ϕ_{sq}) Slip angular frequency $\omega_{sl} = \omega_s - \omega_r$ $\omega_r = P.\ \omega_g$ $\omega_s = 2.\ \pi.f$ Parameters: (R_s, R_r, L_s, L_r, L_m)	$V_{sd} = R_{stator} . I_{sd} - \omega_s . \phi_{sq} + \frac{d\phi_{sq}}{dt}$ $V_{sq} = R_{stator} . I_{sq} + \omega_s . {}_{sd} + \frac{d\phi_{sd}}{dt}$ $0 = R_r . I_{rd} - \omega_{sl} . \phi_{rq} + \frac{d\phi_{rd}}{dt}$ $0 = R_r . I_{rq} + \omega_{sl} . \phi_{rd} + \frac{d\phi_{rq}}{dt}$ $\begin{bmatrix} \$_{sd} \\ \$_{sq} \\ \$_{rd} \\ \$_{rq} \end{bmatrix} = \begin{bmatrix} L_s & 0 & L_m & 0 \\ 0 & L_s & 0 & L_m \\ L_m & 0 & L_r & 0 \\ 0 & L_m & 0 & L_r \end{bmatrix} . \begin{bmatrix} I_{sd} \\ I_{sq} \\ I_{rd} \\ I_{rq} \end{bmatrix}$ $T_{em} = \frac{3}{2} P . \left(\$_{sd} . I_{sq} - \$_{sq} . I_{sd}\right)$
FOC (field-oriented control)	Current, torque and flux references I_{sd-ref}, I_{sq-ref} T_{em-ref}, ϕ_{r-ref}	PI on flux gives I_{sd-ref}: $e = \phi_{r-ref} - \phi_r$ PI on torque/speed or power gives I_{sq-ref}

Table 8.13 (continued)

Sub-system	Main variables	Equations
Three-phase rectifier → DC bus	Phase voltage amplitude (V_m) RMS line-to-line voltage (V_{LL}) Angular frequency (ω) DC voltage (V_{dc}), The diode forward voltage drop V_d Input power (P_{in})	The three-phase voltages at the rectifier input are: $\begin{cases} V_a = V_m \sin(\omega t) \\ V_b = V_m \sin(\omega t - 2\pi/3) \\ V_b = V_m \sin(\omega t + 2\pi/3) \end{cases}$ $V_m = \sqrt{2}\,.V_{LL}/\sqrt{3}$ Average (diode bridge): $V_{dc} = \frac{3\sqrt{2}}{\pi} V_{LL} - 2\,.V_d$ $C\frac{dV_{dc}}{dt} = \frac{P_{in} - P_{out} - P_{losses}}{V_{dc}}$
DC-bus voltage PI	Error $e_v = V_{dc-ref} - V_{dc}$	PI controller for the DC-bus voltage: $u_v(t) = K_{p-dc}\,.e_v(t) + K_{I-dc}\int_0^t e_v(t)\,.dt$
Flywheel energy storage	Angular speed (ω_{FESS}) Inertia (J_{fFESS}) Converter torque applied to the shaft: (T_{conv}) Mechanical power at the shaft: (P_{FESS}) Viscous friction (B_f) Coulomb friction term (T_{cou})	Energy: $E_{FESS} = \frac{1}{2}\,.J_{FESS}\,.\,\omega_{FESS}^2$ Speed limits: $\omega_{FESS-min} \leq \omega_{FESS} \leq \omega_{FESS-max}$ Choose (J_{FESS}) $J_{FESS} = \frac{2\,.E_f}{\omega_{FESS-\max}^2 - \omega_{FESS-\min}^2}$ Dynamics: $J_{FESS}\frac{d\omega_{FESS}}{dt} = T_{conv} - J_{FESS}\,.\,\omega_{FESS} - T_{cou}$ Power: $P_{FESS} = \omega_{FESS}.\,T_{conv}$
Bidirectional DC/DC	Inductor current (i_L) Duty (D_{cycle}) Flywheel DC link (V_f) Capacitance C	$L\frac{dI_L}{dt} = V_{dc}\,.D_{cycle} - V_F(1 - D_{cycle})$ $C\frac{dV_f}{dt} = I_L - I_f$ $P_f = V_f.\,I_f$
Inverter + filter → load	Currents (I_d, I_q) Voltage (V_d, V_q) Filter inductance L Filter resistance R $V_{inv,d}, V_{inv,q}$: Inverter output voltages in d-q frame $V_{load,d}, V_{load,q}$: Load voltages in d-q frame ω: Electrical angular frequency of the rotating reference frame.	Filter (L): $L\frac{dI_d}{dt} = V_{inv,d} - V_{load,d} - R\,.I_d + \omega\,.L\,.I_q$ $L\frac{dI_q}{dt} = V_{inv,q} - V_{load,q} - R\,.I_q + \omega\,.L\,.I_d$ Power: $P_{AC} = \frac{3}{2}(V_d\,.I_d + V_q\,.q)$

(continued)

Table 8.13 (continued)

Sub-system	Main variables	Equations
MPPT	For example TSR: Speed optimal (ω_{Tb}^{opt}) (or torque (T_{opt}))	Optimal speed: $\omega_{Tb}^{opt} = \lambda \,.\, \frac{V_{wind}}{R_{Tb}}$ Optimal torque: $T_{opt} = K_{opt} \,.\, \omega_{Tb}^{2}$ Optimal-torque MPPT coefficient: $K_{opt} = \frac{1}{2}\frac{C_{P-\max} \,.\, \varphi_{Air} \,.\, A_{Tb} \,.\, R_{Tb}^{3}}{\lambda_{opt}^{3}}$

8.2.7 *Wind Turbine with Batteries/Supercapacitors*

Grid access is limited or intermittent in many isolated areas, including agricultural zones, islands, and rural areas. A steady, clean power source can be produced via a hybrid microgrid that combines a wind turbine, battery storage, and a flywheel technology. For example, a 50 kW wind turbine, a 100 kWh lithium-ion battery, and a 20 kWh flywheel system controlled by an energy management control could possibly be used in a remote Mediterranean coastal area (Fig. 8.5). Despite wind fluctuations, this setup would ensure steady power supply and sustainably providing the energy needs of the area. The different parameters are given in Table 8.14.

The wind turbine serves as the main renewable energy source in this hybrid system, using wind energy to generate energy. Stable voltage and frequency is ensured by the flywheel's instantaneous response to sudden fluctuations in wind speed. In the meantime, the battery bank gives electric power when the wind is insufficient or non-existent and stores extra energy during strong wind conditions.

8.2.7.1 Sizing of Studied System

- **Sizing of the wind turbine:**

The instantaneous mechanical power available at the rotor is [13–18].

$$P_{wind} = \left(\frac{1}{2}\right) \mathrm{P}_{air} \,.\, A_{tb} \,.\, V_{wind}^{3} \tag{8.25}$$

To size the rotor area A for a given rated power P_{rated} at a chosen reference wind speed v_{ref} and chosen C_p, we will have:

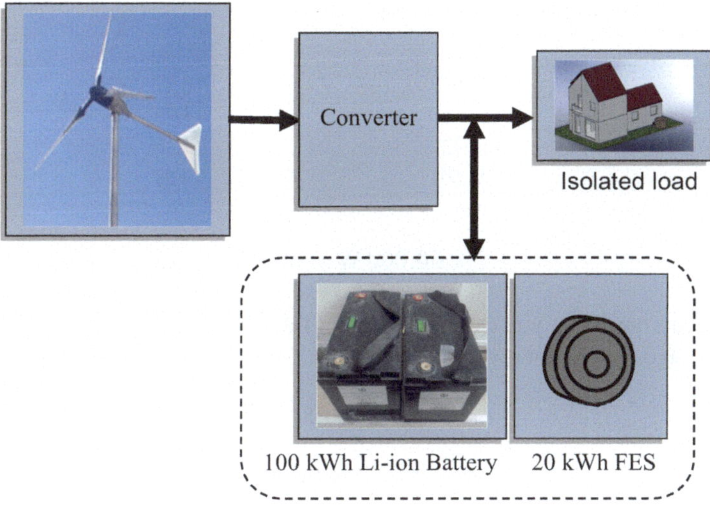

Fig. 8.5 Example representation of WTb with battery and SCs

Table 8.14 Wind turbine parameters

Parameter	Symbol	Value
Wind turbine rated power	P_{rated}	50 kW
Battery nominal energy	$E_{batt,nom}$	100 kWh
Flywheel usable energy	E_{FESS}	20 kWh
DC bus voltage	V_{dc}	400 V
Battery depth of discharge	DOD	0.80
Converter & safety design margin	–	25% (currents/powers × 1.25)
Power coefficient or instantaneous aerodynamic efficiency	C_p	0.40
Air density	ρ_{air}	1.225 kg·m^{-3}
Reference wind speed	V_{ref}	12 m/s

$$A = \frac{2 \,.\, P_{rated}}{\rho_{air} \,.\, C_p \,.\, V_{wind-ref}^3} \tag{8.26}$$

$$D = \sqrt{\frac{4 \,.\, A}{\pi}} \tag{8.27}$$

So, in this application, we obtain:

$$A \approx 118.132\,\mathrm{m}^2 \text{ and } \mathrm{D} \approx 12.26\,\mathrm{m}$$

You can verify

Table 8.15 Calculation of D for different Vref

V_{ref} (m/s)	A (m^2)	D (m)
10	204.082	16.12
11	153.360	13.98
12	118.132	12.26
13	92.891	10.88

$$P \approx 50{,}000\ \text{W} = 50\,kW.$$

If you already have V_{ref}, you can calculate D (Table 8.15):

The average annual energy from turbine is:

$$E_{year} = P_{rated} \times 8760 \tag{8.28}$$

$$E_{year} = 50 \times 8760 = 43800\,kWh\,/\,year$$

And daily average generation is:

$$E_{day} = \frac{E_{year}}{365} = 1200\,kWh\,/\,day$$

Thus, the average power available is:

$$P_{ave} = \frac{1200}{24} = 50\,kW$$

- Sizing of the battery storage:

The usable energy is:

$$E_{batt,usable} = E_{batt,nom} \times DOD \tag{8.29}$$

$$E_{batt,usable} = 100 \times 0.8 = 80\,kWh$$

The nominal capacity is:

$$C_{batt} = \frac{E_{batt,nom}}{V_{dc}} \tag{8.30}$$

$$C_{batt} = \frac{100000}{400} = 250\,Ah$$

Maximum continuous battery current for full turbine power can be calculated as:

$$I_{batt} = \frac{P_{\max}}{V_{dc}} \quad (8.31)$$

$$I_{batt} = \frac{50000}{400} = 125\,A$$

And if we add design margin (×1.25):

$$I_{design} = 1.25 \times I_{batt} \quad (8.32)$$

$$I_{design} = 156.25\text{ A}$$

- Flywheel sizing

The flywheel's power supply duration (t_{supply}) during discharge is determined by its energy capacity (E_{FESS}).

The flywheel stored energy is:

$$E_{FESS} = \frac{1}{2} J_{FESS} \times \omega_{FESS}^{2} \quad (8.33)$$

$$E_{FESS} = 10\,kWh = 10 \times 3.6\,MJ = 36\,MJ$$

For example: If $N = 30{,}000$ rpm, $\omega = 2\pi.(500) \approx 3142$ rad/s.

$$\mathrm{J}_{FESS} = \frac{2.\mathrm{E}_{FESS}}{\omega_{FESS}^{2}} \quad (8.34)$$

$$J = \frac{2 \times 36}{314^{2}} = 7.29\,kg\,/\,m2$$

If the flywheel provides a brief interruption:

$$t_{supply} = \frac{E_{FESS}}{P_{FESS}} \quad (8.35)$$

Considering this relationship is crucial for understanding how rapid power smoothing and transient response in hybrid renewable systems.

For example, for smoothing peaks of 50 kW, we will have the different values of t_{supply} (Table 8.16):

Table 8.16 Relationship between flywheel energy capacity and supply duration for a 50 kW system

E_{FESS} (kWh)	P_{FESS} (kW)	t_{supply} (s)	t_{supply} (min)
0.1	50	7.2	0.12
0.5	50	36	0.6
1	50	72	1.2
2	50	144	2.4
5	50	360	6.0
10	50	720	12.0
20	50	1440	24.0

8.2.7.2 Converter and Inverter Sizing

At least a continuous power rating of Prated or the expected maximum discharge power (e.g., 50 kW) and the margin of design is from 1.2 to 1.5. So, in this application, select a converter rated between 60 and 75 kW for continuous operation. For Flywheel, the peak power must be managed for expected transients (2× average smoothing power) must be selected. For example, design for short-term, 100–150 kW peak power capability if a flywheel needs to absorb or deliver 100 kW. Also, if the AC village grid is supply by this system, the inverter should be rated for peak load (average peak of 1.2×). So, inverter ≥75 kW is the best option if peak loads are up to 60 kW.

8.3 Application of Chap. 5

8.3.1 PV/Batteries/FCs System

The system can be represented in Fig. 8.6. An autonomous load is directly supplied by the DC power produced by the PV panels. A DC/DC converter provides extra electrical power to the electrolyser when PV generation exceeds load demand. Hydrogen (H_2) is created by the electrolyser using this additional power and is kept in the H_2 tank for later use [19].

In this application, it is required to evaluate compare three operating scenarios for an off-grid PV system that provides a daily electricity load in this application. Under base conditions, a 0.64 kWp PV array with an average load of 2.808 kWh/day (Fig. 8.7) produces 3.20 kWh/day during five peak solar hours (*PSH* = 5 h). Assuming a constant load, the result is 0.585 kWh during the day (5 h) and 2.223 kWh during the night. The objective is to assess how the system can use various energy-storage arrangements to maintain supply during the night or over several days of autonomy.

Thus, Table 8.17 provides a summary of the various calculations.

Scenario 1 depicts a PV + PEMFC system without battery storage. Here, the fuel cell operates at night to supply the entire 2.223 kWh demand when the PV is not in

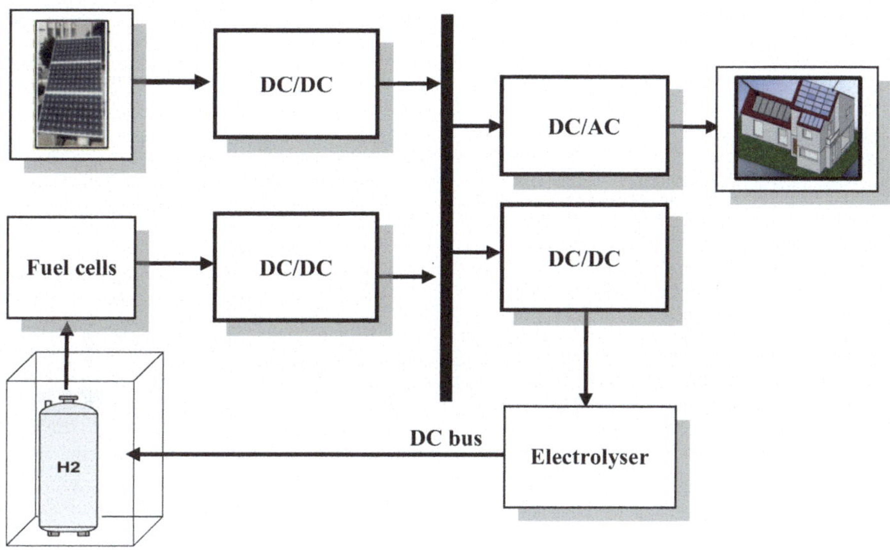

Fig. 8.6 PV system with FC storage supplying autonomous load

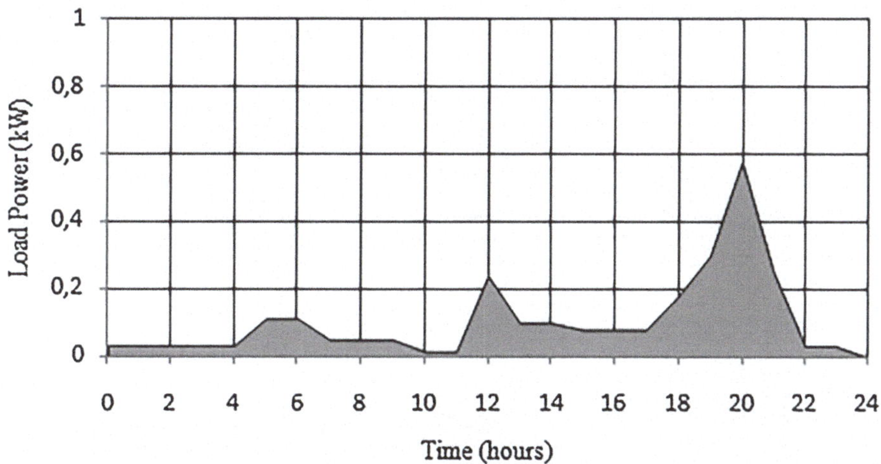

Fig. 8.7 Chosen load profile power

use. The hydrogen lower heating value (33.3 kWh/kg) and cell efficiency (50%) allow us to calculate the fuel cell's electrical power output (about 117 W) and related hydrogen consumption.

Scenario 2 explores at a PV + battery system without a fuel cell. The battery's required capacity, which supplies the 2.223 kWh night load, is determined by the intended autonomy period. We can determine the nominal battery energy for one, three, and seven days of autonomy while accounting for the depth of discharge (0.8) and discharge efficiency (0.8).

Table 8.17 Different scenarios sizing in PV/Battery/PEMFC

Cases/scenarios				Calculations			
Scenario1	PV	PEMFC	Battery	$E_{PV}=P_{PV}.PSH$ E_{load} *(given)* $DP = E_{load} - E_{PV}$	$E_{PV} = 3.20$ kWh/day $E_{load} = 2.808$ kWh/day $DP = 0.392$ kWh/day.		
				$E_{day} = E_{load}.(PSH/24)$	$E_{day} = 2.808 \times (5/24) = 0.585$ kWh		
				$E_{night} = E_{load} - E_{day}$	$E_{night} = 2.808–0.585 = 2.223$ kWh/day		
				Night length: 24 – PSH	*Night length* = 24–5 = 19		
				$P_{FC,night} = E_{def}/19$	$P_{FC,night} = 2.223/19 = 0.117$ kW $\approx$ 117 W		
	X	**X**		$E_{H2} = E_{def}/\eta_{FC} = E_{def}/0.5$	$E_{H2} = 2.223/0.5 = 4.446$ kWh/day		
				$m_{H2} = E_{H2}/LHV_{H2}$	$m_{H2} = 4.446/33.3 = 0.1335$ kg/day		
Scenario2	**X**		X	$E_{batt,req} = N \cdot E_{night}/(DODx\eta_{batt})$ $= N \cdot 2.223/(0.8x0.8)$	N_{aut} (days)	$E_{batt,req}$ (kWh)	
					1	3.473	
					3	10.42	
					7	24.314	
Scenario3	**X**	**X**	**X**	$P_{FC} = 0.117 \times f$	$P_{FC} = 0.117 \times 0.5 = 0.0585$ kW		
				$E_{batt,nom} = N_{aut} \cdot (1-f)E_{night}/(DODx\eta_{batt})$ $=N_{aut} \cdot (1–0.5)E_{night}/(0.8x0.8)$ $m_{H2,night} = f.E_{FC,night}/LHV_{H2} \cdot \eta_{FC}$	N_{aut} (days)	$E_{batt,nom}$ (kWh)	$m_{H2,night}$ Over night (kg)
					1	$=1 \times (1–0.5) \times 3.473$ $=1.737$	1×0.06675 $=0.0668$
					3	$=3 \times 0.5 \times 3.473$ $=5.210$	3×0.06675 $=0.200$
					7	$=7 \times 0.5 \times 3.473$ $=12.156$	7×0.06675 $=0.467$

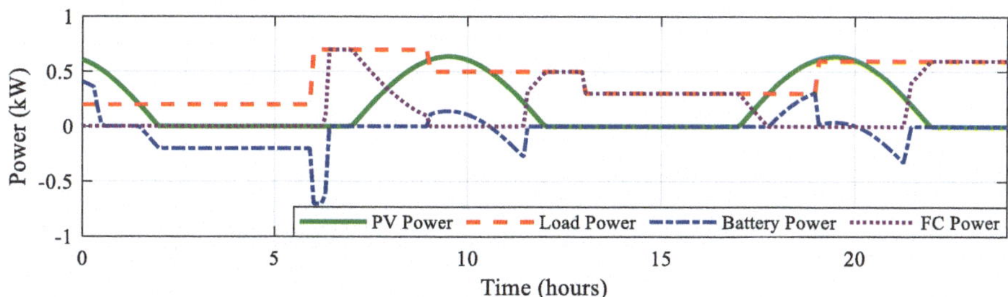

Fig. 8.8 Evolution of the different powers variation

Scenario 3 introduces a compensated hybrid mode that integrates PV, battery, and PEMFC. Both the fuel cell and the battery contribute to the night load in this configuration. Assuming an energy-sharing factor ($f_{sharing} = 0.5$), the battery provides half of the night load while the fuel cell provides the remaining half. For the same autonomous days (1, 3, and 7 days), the required battery capacity, average fuel-cell power, and total hydrogen consumption are recalculated. This hybrid example demonstrates how combining chemical and electrochemical storage can reduce the size of each subsystem while maintaining energy availability during extended cloudy periods.

The PV system with fuel cell storage model and batteries is coded under Matlab and the different power evolution are displayed in Fig. 8.8.

8.3.2 PV/Wind Turbine System with SC Storage

We propose an application with the different SC parameters (Table 8.18):

8.3.2.1 Sizing Cell Max Voltage

For each cell, we will have:

$$E_{cell} = 3.0\,Wh = 3.0 \times 3600 = 10{,}800\,\text{J}.$$

If $C_{cell} = 3000$ F, then cell voltage at that energy is:

$$V_{cell} = \sqrt{\frac{2\,.\,E_{cell}}{C_{cell}}} \tag{8.36}$$

$$\text{V}_{\text{cell}} = \sqrt{\frac{2.10800}{3000}} = \sqrt{7.2} = 2.683\,V$$

Table 8.18 The different parameters of SC

Parameter	Description	Value
C_N	Pack nominal	165 F
ESR_{DC}	Pack ESR	60 mΩ
IR_{DC}	Standard test current	100 A
V_N	Pack nominal voltage	48 V
E_{sc}	Pack energy	53 Wh
V_{max}	Pack max	51 V
I_{max}	Peak current capability	1900 A
V_{series}	Application/system voltage target	750 V
C_{cells}	Single cell capacitance	3000 F
$E_{sc-cell}$	Single cell energy	3.0 Wh
N_{cells}	Cells in series	18

Thus the pack nominal voltage is:

$$V_{pack} = N \,.\, V_{cell} \tag{8.37}$$

$V_{pack} = 18\text{x}2.683 = 48.29\ V$ Which matches with $V_N = 48$ V

8.3.2.2 Sizing Pack Capacitance

We have:

$$C_{pack} = \frac{C_{cell}}{N} \tag{8.38}$$

$C_{pack} = \frac{3000}{18} = 166.67$ F Which corresponds to the given $C_N = 165$ F

8.3.2.3 Pack Energy Sizing

We have:

$$E_{pack} = N \,.\, E_{cell} \tag{8.39}$$

$E_{pack} = 18\text{x}3 = 54$ Wh Which is close to the given pack energy value $E_{sc} = 53$ Wh
And we can check with the following equation:

$$E_{pack} = \frac{1}{2} \,.\, C_{pack} \,.\, V_{pack}^2 \tag{8.40}$$

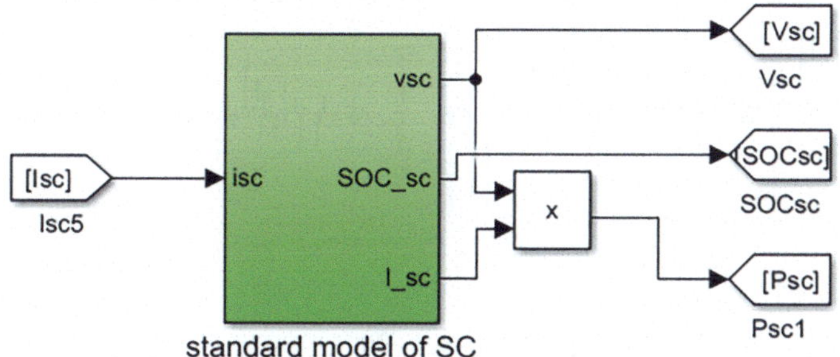

Fig. 8.9 SC model in Matlab/Simulink

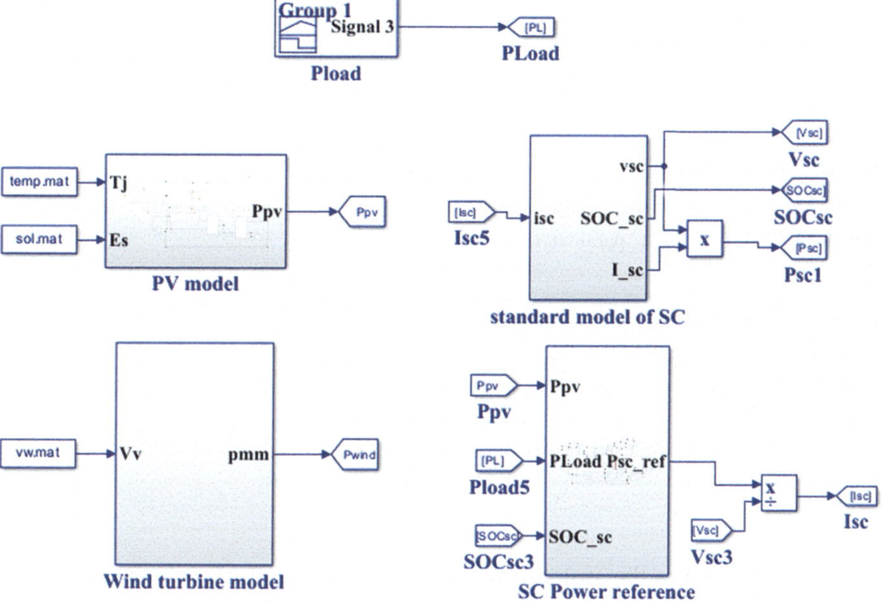

Fig. 8.10 Application of RES with SC storage

$$E_{pack} = \left(\frac{1}{2} . (166.67).(48.29^2) \right) / 3600 = 53.98\,Wh \approx 54\,Wh$$

8.3.2.4 Modelling Under Matlab/Simulink

The model in Matlab/Simulink can be depicted as (Fig. 8.9) based on the various equations.

A application in Matlab/Simulink is made to a renewable system (PV with wind turbine) with SC storage (Fig. 8.10). The dynamic performance of the hybrid RES combining PV, wind turbine, and SC storage is shown via simulation results

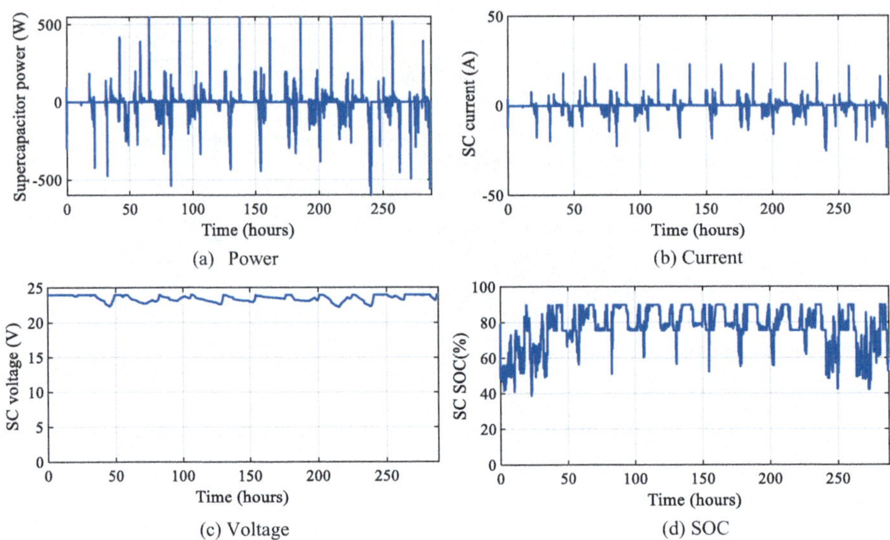

Fig. 8.11 Simulation results using SC storage in RES. (**a**) Power. (**b**) Current. (**c**) Voltage. (**d**) SOC

Table 8.19 PEMFC and SC parameters

Parameters	Values
Vacuum voltage	45 V
Nominal voltage	26 V
Nominal current	46 A
Maximum current	50 A
Area	50 cm^2
C	2700F
ESR	0.8 mΩ.

(Fig. 8.11). To assess the system's power balance and storage behavior, different wind speed and solar irradiance profiles were simulated. The dynamic performance of the hybrid renewable energy system combining PV, wind turbine, and SC storage is shown via simulation results.

The supercapacitor's ability to compensate for sudden power fluctuations is highlighted by the voltage and current profiles, which show quick reaction during transient load or generation variations. The SOC of SC maintains operation within the required voltage limits by easily switching between the charging and discharging phases. The instantaneous power curves show that the SC absorbs excess power when renewable energy exceeds load demand and discharges to stabilize the DC bus during deficits.

8.3.3 Application of PEMFC/SCs Suppling Electric Vehicle

In this application a PEMFC of 1.2 kW is chosen (Table 8.19).

The PEMFC and SC model is implemented in Simulink (Fig. 8.12).

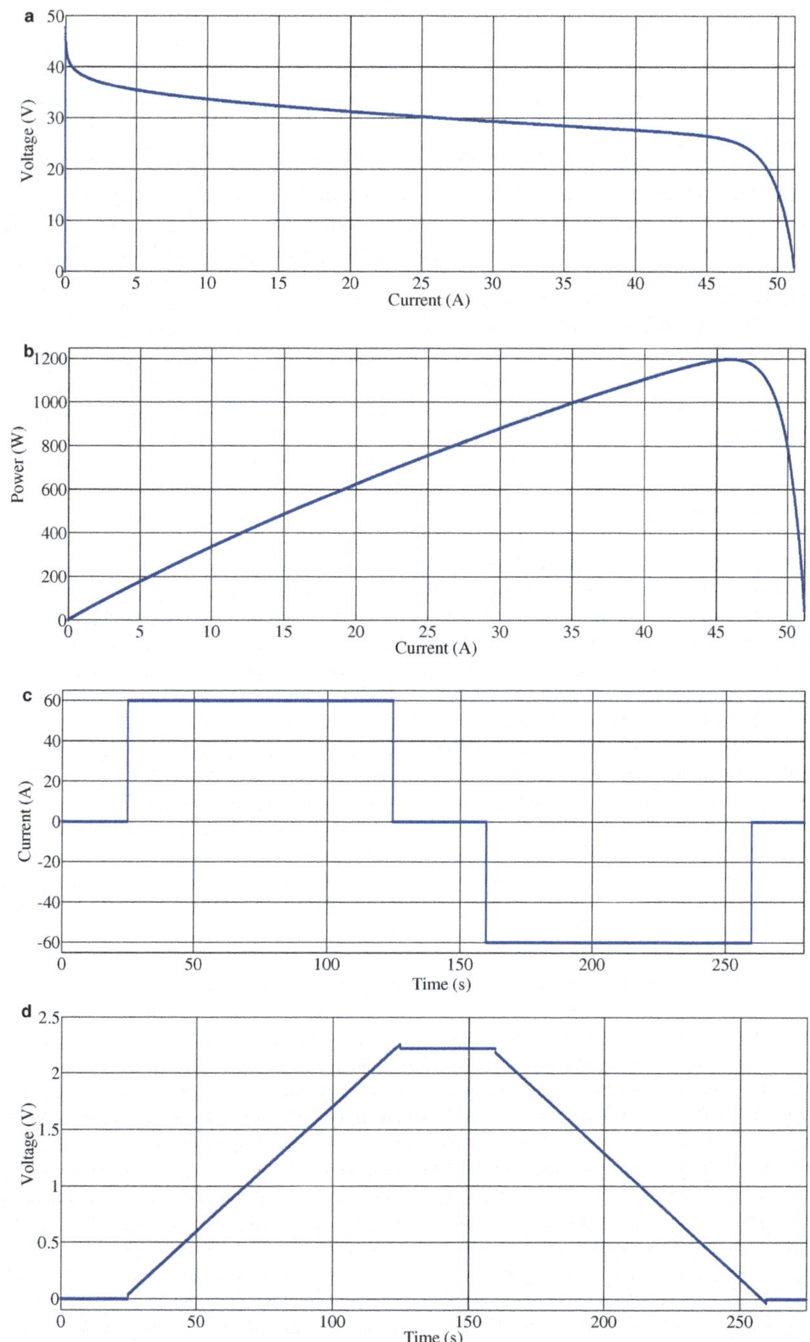

Fig. 8.12 PEMFC and SC simulation results. (**a**) Voltage-Current characteristic of PEMFC. (**b**) Power-current characteristic of PEMFC. (**c**) Applied charge-discharge current waveform to the supercapacitor. (**d**) Voltage response of the supercapacitor for applied current

8.4 Conclusion

This chapter has demonstrated diverse applications of hybrid RESs integrating batteries, fuel cells, and supercapacitors with solar and wind generation. The simulation results confirm that hybrid architectures significantly improve energy reliability, reduce intermittency, and ensure optimal power balance under dynamic operating conditions. The coordinated use of storage devices, with batteries providing medium-term energy smoothing, supercapacitors managing rapid transients, and fuel cells supplying steady long-term power, proves essential for maintaining both system stability and efficiency.

References

1. Zaouche F, Rekioua D, Gaubert J-P, Mokrani Z (2017) Supervision and control strategy for photovoltaic generators with battery storage. Int J Hydrog Energy 42(30):19536–19555. https://doi.org/10.1016/j.ijhydene.2017.06.107
2. Hassani H, Rekioua D, Zaouche F, Bacha S (2019) Supervision of hybrid renewable energy systems. In: 2019 1st International Conference on Sustainable Renewable Energy Systems and Applications (ICSRESA). IEEE, pp 1–5. https://doi.org/10.1109/ICSRESA49121.2019.9182478
3. Rekioua D, Zaouche F, Hassani H, Rekioua T, Bacha S (n.d.) Modeling and fuzzy logic control of a stand-alone photovoltaic system with battery storage. Turkish J Electromech Energy 4(1):11–17. https://sloi.org/urn:sl:tjoee41137
4. Serir C, Tadjine K, Bensmail S, Rekioua D, Belkaid A, Hadji S, Colak I (2024) Smart energy management control based on fuzzy logic controller in a standalone photovoltaic/wind system with battery storage. Elect Power Compon Syst 52(7):1145–1154. https://doi.org/10.1080/15325008.2024.2309631
5. Aissou R, Rekioua T, Rekioua D, Tounzi A (2016) Application of nonlinear predictive control for charging the battery using wind energy with permanent magnet synchronous generator. Int J Hydrog Energy 41(45):20964–20973. https://doi.org/10.1016/j.ijhydene.2016.05.249
6. Rekioua D, Mokrani Z, Kakouche K et al (2024) Coordinated power management strategy for reliable hybridization of multi-source systems using hybrid MPPT algorithms. Sci Rep 14:10267. https://doi.org/10.1038/s41598-024-60116-4
7. Kakouche K, Oubelaid A, Mezani S, Rekioua D, Rekioua T (2023) Different control techniques of permanent magnet synchronous motor with fuzzy logic for electric vehicles: analysis, modelling, and comparison. Energies 16(7):3116. https://doi.org/10.3390/en16073116
8. Rekioua D (2020) MPPT methods in hybrid renewable energy systems. In: Hybrid renewable energy systems. Springer, pp 79–138. https://doi.org/10.1007/978-3-030-34021-6_3
9. Rekioua D, Kakouche K, Babqı A, Mokrani Z, Oubelaid A, Rekioua T, Azil A, Ali E, Alaboudy AHK, Abdelwahab SAM (2023) Optimized power management approach for photovoltaic systems with hybrid battery–supercapacitor storage. Sustainability 15(19):14066. https://doi.org/10.3390/su151914066
10. Rekioua D, Mokrani Z, Rekioua T (n.d.) Control of fuel cells–electric vehicle based on direct torque control. Turkish J Electromech Energy 3(2)
11. Abderezzak B, Rekioua D, Binns R, Busawon K, Hinaje M, Douine B et al (2020) Technical feasibility assessment of a PEM fuel cell refrigerator system. Int J Hydrog Energy 45(19):11211–11219

12. Rekioua D, Bensmail S, Bettar N (2014) Development of hybrid photovoltaic–fuel cell system for stand-alone application. Int J Hydrog Energy 39(3):1604–1611. https://doi.org/10.1016/j.ijhydene.2013.03.040
13. Tadjine K, Rekioua D, Belaid S, Rekioua T, Logerais PO (2022) Design, modeling and optimization of hybrid photovoltaic/wind turbine system with battery storage: application to water pumping. Math Model Eng Prob 9(3):655–667. https://doi.org/10.18280/mmep.090312
14. Aissou R, Rekioua T, Rekioua D, Tounzi A (2016) Robust nonlinear predictive control of permanent magnet synchronous generator turbine using Dspace hardware. Int J Hydrog Energy 41:21047–21056
15. Belaid S, Rekioua D, Oubelaid A, Ziane D, Rekioua T (2022) Power management control and optimization of a wind turbine with battery storage system. J Energy Storage 45:103613. https://doi.org/10.1016/j.est.2021.103613
16. Rekioua D, Mezzai N, Mokrani Z et al (2024) Effective optimal control of a wind turbine system with hybrid energy storage and hybrid MPPT approach. Sci Rep 14:30013. https://doi.org/10.1038/s41598-024-78847-9
17. Rekioua D (2023) Energy storage systems for photovoltaic and wind systems: a review. Energies 16(9):3893. https://doi.org/10.3390/en16093893
18. Rekioua D (2020) Power electronics in hybrid renewable energies systems. In: Hybrid renewable energy systems. Springer, pp 39–77. https://doi.org/10.1007/978-3-030-34021-6_2
19. Rekioua D, Roumila Z, Rekioua T (2008) Étude d'une centrale hybride photovoltaïque–éolien–diesel. J Renew Energ 11(4):623–633. https://doi.org/10.54966/jreen.v11i4.112
20. Rekioua D (2024) Wind power electric systems: modeling, simulation and control. Springer. https://doi.org/10.1007/978-3-031-52883
21. Rekioua D, Matagne E (2012) Optimization of photovoltaic power systems: modeling, simulation and control. Springer. https://doi.org/10.1007/978-1-4471-2403-05
22. Rahrah K, Rekioua D, Rekioua T, Bacha S (2015) Photovoltaic pumping system in Bejaia climate with battery storage. Int J Hydrog Energy 40(39):13665–13675. https://doi.org/10.1016/j.ijhydene.2015.04.048
23. Ould-Amrouche S, Rekioua D, Hamidat A (2010) Modelling photovoltaic water pumping systems and evaluation of their CO_2 emissions mitigation potential. Appl Energy 87(11):3451–3459. https://doi.org/10.1016/j.apenergy.2010.05.021
24. Adjati A, Rekioua T, Rekioua D (2021) Use of the dual stator induction machine in photovoltaic–wind hybrid pumping. Journal Européen des Systèmes Automatisés 54(1):115–124. https://doi.org/10.18280/jesa.540113

Conclusion

Energy storage systems (ESSs) have become fundamental to the advancement, reliability, and large-scale deployment of renewable energy systems. Effective integration of storage technologies is essential for maintaining system stability, maximizing performance, and mitigating the inherent intermittency of renewable resources. The purpose of this book has been to provide readers with a comprehensive and structured understanding of ESSs, spanning basic principles, modeling approaches, control strategies, and advanced applications.

This book serves as a valuable reference for students, academics, engineers, and practitioners working in renewable energy, smart grids, and energy storage technologies. By combining theoretical foundations with practical methodologies, it aims to equip readers with the knowledge required to design, analyze, and manage modern energy storage solutions.

Although significant technological progress has been achieved in recent years, numerous challenges remain and many opportunities for innovation still exist. Future research will likely be driven by advancements in hybrid storage architectures, AI-driven and predictive energy management, next-generation battery chemistries, and techno-economic and life-cycle assessment models, as well as other strategic directions.

To all readers: it is our hope that this book motivates and inspires you to explore novel ideas and research directions and to advance the frontiers of energy storage and renewable energy systems.

D. Rekioua, *Energy Storage for Renewable Energy Systems*, Green Energy and Technology, https://doi.org/10.1007/978-3-032-19589-0